# Inspection of Concrete Elements Reinforced or Strengthened with Fiber-Reinforced Polymers

Fiber-Reinforced Polymers (FRPs) offer higher durability and better performance than steel in strengthening and repairing existing structures, whether used as externally bonded systems or as internal reinforcement for concrete elements, and can increase the life cycle of concrete elements. Nevertheless, FRP-reinforced or strengthened concrete (FRP-RSC) elements need to be inspected regularly for effective maintenance and to ensure safety.

This book addresses the absence of inspection guide and procedures for FRP-RSC. It reviews serviceability considerations along with specification and standards for design and installation. It identifies potential damage and anomalies attributed to FRP-RSC elements and evaluates inspection methods. It also proposes solutions for improving the detectability of FRP defects such as non-destructive testing methods. It offers a guide for in-service inspection of FRP-RSC elements, condition rating tables, examples, and guidance for load rating.

**Armin Mehrabi** is an associate professor at Florida International University (FIU), USA, with three decades of experience in evaluation, inspection, NDT, rehabilitation, and maintenance of bridges.

**Antonio Nanni** is a research professor at the University of Miami (UM), USA, with over four decades of academic and professional experience in design, new construction, and repair of concrete structures using composites.

**Pranit Malla** is a bridge engineer at AtkinsRéalis and a Postdoctoral Associate at Florida International University, specializing in the inspection and evaluation of bridges with FRP applications.

**Seyed Saman Khedmatgozar Dolati** is a post-doctoral associate at FIU and bridge engineer at HDR, specializing in innovative design, construction, and inspection. He is a recipient of the prestigious 2024 FIU Presidential Award.

**Jesus D. Ortiz** is a forensic and renewal engineer at Thornton Tomasetti who specializes in building inspections and evaluations to determine the cause, origin, and extent of damage across a wide variety of events.

# Inspection of Concrete Elements Reinforced or Strengthened with Fiber-Reinforced Polymers

Armin Mehrabi, Antonio Nanni, Pranit Malla,
Seyed Saman Khedmatgozar Dolati, and
Jesus D. Ortiz

CRC Press
Taylor & Francis Group
Boca Raton London New York

CRC Press is an imprint of the
Taylor & Francis Group, an **informa** business

First edition published 2026
by CRC Press
4 Park Square, Milton Park, Abingdon, Oxon, OX14 4RN

and by CRC Press
2385 NW Executive Center Drive, Suite 320, Boca Raton FL 33431



*British Library Cataloguing-in-Publication Data*
A catalogue record for this book is available from the British Library

ISBN: 9781041086420 (hbk)
ISBN: 9781041087908 (pbk)
ISBN: 9781003646969 (ebk)

DOI: 10.1201/9781003646969

Typeset in Times
by Deanta Global Publishing Services, Chennai, India

# Contents

# 1 Background on Materials and Installation

## 1.1 INTRODUCTION (HISTORICAL PERSPECTIVE)

The 1930s can be regarded as an important decade in the history of fiber-reinforced polymer (FRP) composites, as the components (fibers and matrices) required for manufacturing the FRP composites were beginning to come together during this period. In 1932, the mass production of glass fibers was accidentally invented by the Owens Glass Company when a jet of compressed air was directed at a stream of molten glass. Later in 1935, the Owens Glass Company joined with the Corning Company to produce the first patented fiberglass [1]. Subsequently, in 1938, DuPont patented the first suitable polyester resin to be used for the preparation of fiber composites [1]. Hence, the 1930s can be considered as the beginning of the FRP industry since the fibers and resins developed during that time are still the dominant choice as the components of FRP composites manufactured today.

During World War II, in the 1940s, the FRP industry moved from laboratory research to actual production. The wartime needs for lightweight and strong composites led to several of the greatest advancements in the FRP industry. The first fiberglass laminates with low-pressure polyester resins were produced by Pittsburgh Plate Glass Company for aircraft, boat, and automobile parts in 1942 [2]. Over the next 20 years, during the 1950s and 1960s, the advancements and innovations in the manufacturing technologies of the FRP composites continued, but their application was still predominantly limited to the military and aerospace industries.

One of the first civil engineering applications of FRP composites was a dome constructed in Benghazi, Libya, in 1968; similarly, the first FRP bridge (pedestrian) was built in Israel in 1975 [3]. The early research in the application of FRP composites for strengthening and repair of various structural members was concentrated in Japan, Switzerland, and Germany since the late 1970s and early 1980s [4]. The use of FRP materials for retrofitting concrete structures was first documented in Germany in 1978 [5]. Research on exploring the use of FRP composites in the repair and retrofitting of civil infrastructure started in Japan in the early 1980s. In Japan, the first composite repair application was performed in 1984, when cracks in railway bridge piers were repaired with carbon fiber sheets [4]. One of the first field applications of FRP strengthening in Europe was on the Ibach Bridge in Lucerne, Switzerland, in 1991 [6]. The United States and Canada joined efforts in research and applications of strengthening with FRP composites by the late 1980s and lagged almost a decade behind Japan and Europe. A precast post-tensioned bridge built in 1991 in Rapid City, South Dakota, that used three different materials, GFRP, CFRP, and

DOI: 10.1201/9781003646969-1

steel cables, to prestress the deck slab was the first of its kind in the United States [7, 8]. Since then, various research studies in North America have been conducted to investigate the possible applications of FRP composites in civil engineering, leading to widespread application from the start of the twenty-first century.

## 1.2 FIBER-REINFORCED POLYMER (FRP) COMPOSITES

FRPs are composite materials consisting of reinforcing fibers embedded/impregnated in polymeric resins. There are mainly three types of fibers used for bridge construction: glass, carbon, and aramid fibers. In recent years, basalt fibers have also been used increasingly in the construction industry. While these fibers provide the load-bearing action, the polymeric matrix (or polymeric resin) allows for the creation of a desired geometry and transfers the forces to the fibers. The matrix also prevents fibers from buckling when in compression and protects them from degrading environments such as humidity [9]. Similarly, the basic types of polymers that can be used as matrices are thermoplastics and thermosets (e.g., epoxies, polyesters, and vinyl esters), with thermosets being used more commonly. The main difference between these two types of resins is that thermoplastics can be reheated for bending without much damage to the matrix, while thermosets cannot be bent under heat once cured (an in-depth explanation is provided later in this chapter). Different additives and fillers can also be mixed with polymeric resin to improve performance, tailor composite performance, and reduce costs [10]. Figure 1.1 shows the composition of FRP composites typically used in civil engineering applications.

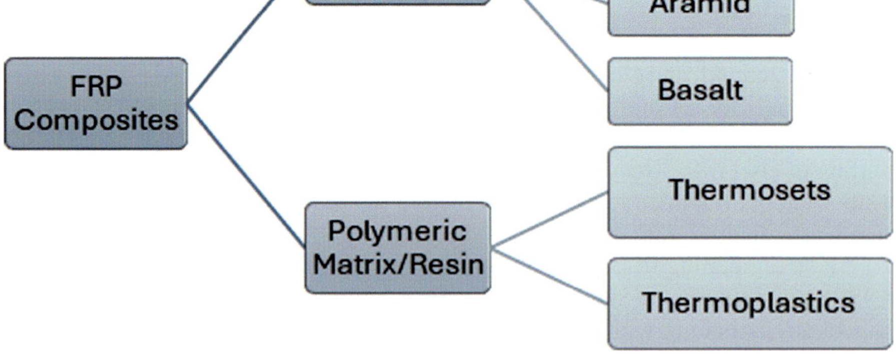


**FIGURE 1.1** Composition of FRP composites used in civil engineering applications

### 1.2.1 Reinforcing Fibers

Fibers are the reinforcing component of composite materials. They have a high aspect ratio, i.e., length to diameter ratio, of usually greater than 100 (very large for continuous fibers), with a diameter in the order of 10 μm [11]. The reinforcing fibers have exceptionally high specific stiffness and strength; however, they may be brittle and vulnerable to surrounding harsh environments. Even simple handling or friction can easily damage them. Thus, for the high strength application of these fibers, they need to be coated with materials that can provide the protective layer and further help transfer stresses between them. This is achieved by impregnating the reinforcing fibers with a polymeric matrix.

#### 1.2.1.1 Glass Fibers

Glass fibers are the most commonly used fiber type in FRP composites as they are more economical compared to carbon and aramid fibers. Moreover, glass fibers have higher ultimate strain compared to the other two reinforcing fibers. Lower tensile modulus, higher specific gravity, and lower fatigue resistance compared to carbon and aramid fibers, as well as their vulnerability to abrasion, are some of their disadvantages [12].

Glass fibers comprise silica-based compounds with several metal oxides that can be tailored for desired material properties such as corrosion resistance and wettability [4]. Considering all applications, E-glass (electrical glass) and S-glass (structural glass) are the two most commonly used glass fibers in FRP composites. The glass fibers with a calcium alumina borosilicate composition are called E-glass fibers, and those with an aluminosilicate composition are called S-glass fibers [13]. E-glass fibers are the most commonly used glass fibers in the construction industry for general-purpose structural applications and the retrofitting of structural components. In comparison, these fibers have good heat and electrical resistance, low susceptibility to moisture, and low cost. On the other hand, S-glasses are more expensive than E-glass, but they have higher strength, stiffness, and corrosion resistance than E-glass. S-glass fibers have typically 40% greater strength at room temperature than E-glass [11]. The tensile properties of these glass fibers used in FRP composites are given in Table 1.1.

E-glass fibers, however, are more vulnerable to acidic and alkaline environments as compared to other available special-purpose glass fibers such as C, R, and

**TABLE 1.1**
**Typical tensile properties of glass fibers used in FRP systems [5]**

| | Elastic modulus | | Ultimate strength | | |
|---|---|---|---|---|---|
| Type | $10^3$ ksi | GPa | ksi | MPa | Rupture strain, minimum, % |
| E-glass | 10 to 10.5 | 69 to 72 | 270 to 390 | 1860 to 2680 | 4.5 |
| S-glass | 12.5 to 13 | 86 to 90 | 500 to 700 | 3440 to 4140 | 5.4 |

ECR glass fibers. Hence, ECR glass fibers, one of the special-purpose glass fibers with high corrosion resistance, have gained considerable commercial significance in the market today and are widely used in the field of civil engineering. These chemical-resistant E-glass fibers (ECR glass fibers) were developed after continuous improvements in formulations for E-glass fibers [14]. The ECR glass fibers are hence the modified E-glass fibers that offer enhanced long-term acid resistance and short-term alkali resistance [15].

#### 1.2.1.2 Carbon Fibers

Carbon fibers are the next widely used reinforcing fibers in FRP composites [12]. They have high specific tensile strength and stiffness, low coefficient of thermal expansion, and good fatigue resistance. However, their disadvantages include high cost, high electrical conductivity, and low impact resistance. Moreover, lower ultimate strain compared to other fibers and potential galvanic corrosion when used in contact with metals might restrict their use in certain applications [4].

Carbon fibers are manufactured by pyrolysis of certain organic/hydrocarbon precursors which could be either a polyacrylonitrile (PAN) precursor or a pitch precursor. Manufacturing carbon fibers from PAN precursors is economical as 50–55% of the fiber mass is retained during the processing/carbonization of PAN fibers into carbon fibers. In addition, the carbon fibers manufactured using PAN precursors have higher tensile strengths than those made from other precursors [11] due to the lack of surface defects in them [13]. The second type of commonly used precursor is called pitch, which is a byproduct of petroleum refining. Manufacturing carbon fibers from pitch precursor has a higher conversion yield of carbon fibers and costs less than PAN precursors. However, these fibers have lower tensile and compressive strength due to the surface irregularities [14].

The stiffness and strength properties of carbon fibers can be modified based on the heat treatment temperature. Hence, by tailoring the manufacturing process, carbon fibers of several specific categories, such as general purpose, high strength, ultra-high strength, high modulus, and ultra-high modulus, can be made. However, gaining higher strength will be at the cost of reducing the modulus of the fibers and vice versa, i.e., improving one material property will result in a reduction in another material property [12]. The typical tensile properties of different categories of available carbon fibers used in FRP systems are given in Table 1.2.

#### 1.2.1.3 Aramid Fibers

Aramid fibers have the lowest specific gravity and the highest tensile strength to weight ratio among other fibers such as glass, carbon, and basalt fibers [12]. Other advantages of aramid fibers include high toughness, resistance to impact damage, and resistance to fatigue. Similarly, these fibers are nonconductive and are resistant to organic solvents. However, high cost, tendency to absorb moisture, and low adhesion to matrix resins are some of its major disadvantages [4]. The aramid fibers are available in two grades of stiffness: (1) general-purpose aramid fibers, which are used in bullet-proof systems, and (2) high-performance aramid fibers, which are used in FRP composites for civil engineering applications [11]. The typical tensile

**TABLE 1.2**
**Typical tensile properties of carbon fibers used in FRP systems [5]**

| | Elastic modulus | | Ultimate strength | | Rupture strain, minimum, % |
|---|---|---|---|---|---|
| **Type** | **$10^3$ ksi** | **GPa** | **ksi** | **MPa** | |
| General purpose | 32 to 34 | 220 to 240 | 300 to 550 | 2050 to 3790 | 1.2 |
| High-strength | 32 to 34 | 220 to 240 | 550 to 700 | 3790 to 4820 | 1.4 |
| Ultra-high-strength | 32 to 34 | 220 to 240 | 700 to 900 | 4820 to 6200 | 1.5 |
| High-modulus | 50 to 75 | 340 to 520 | 250 to 450 | 1720 to 3100 | 0.5 |
| Ultra-high-modulus | 75 to 100 | 520 to 690 | 200 to 350 | 1380 to 2400 | 0.2 |

properties of these two categories of aramid fibers used in FRP systems are given in Table 1.3.

### 1.2.1.4 Basalt Fibers

Recently, basalt fiber-reinforced polymer (BFRP) has been showing some potential with comparable advantages over other FRPs while providing some cost benefit [16, 17]. Basalt fibers are made by drawing fibers from melted basalt rocks and hence are nonhazardous and possibly more environmentally friendly materials. Compared to the commonly used E-glass FRP (GFRP), BFRP may provide higher strength, modulus, and chemical stability at a similar cost [18]. And compared to the conventional steel reinforcements, BFRP has over five times the strength with only about one-third the density [19]. Lopresto et al. (2011) showed better performance for BFRPs in terms of Young's modulus, compressive and bending strength, impact force, and energy as compared to E-glass fibers and hence suggested that it is possible to replace the glass composites with basalt composites in most of the applications [20].

**TABLE 1.3**
**Typical tensile properties of aramid fibers used in FRP systems [5]**

| | Elastic modulus | | Ultimate strength | | Rupture strain, minimum, % |
|---|---|---|---|---|---|
| **Type** | **$10^3$ ksi** | **GPa** | **ksi** | **MPa** | |
| General purpose | 10 to 12 | 69 to 83 | 500 to 600 | 3440 to 4140 | 2.5 |
| High-performance | 16 to 18 | 110 to 124 | 500 to 600 | 3441 to 4140 | 1.6 |

### 1.2.2 Matrices

The main purpose of the matrix is to bind the reinforcing fibers, transfer load to and between the fibers, and to protect the reinforcing fibers from the environment and handling [15]. There are mainly two types of matrices used in FRP composites: thermoplastic and thermosetting matrices.

#### 1.2.2.1 Thermoplastic Matrices

Polymers where molecules are bound together by weak bonds such as hydrogen bonds or Van der Waals bonds are known as thermoplastic matrices. Thermoplastic matrices do not have permanent chemical bonds between the polymer molecules, and their weak bonds can be temporarily broken down via pressure and heat, enabling them to take new forms that can be retained upon elimination of pressure and heat. Therefore, a thermoplastic matrix can be melted and reshaped repeatedly simply by cooling and heating [4]. Some of the advantages of a thermoplastic matrix are fracture resistance, high impact strength, short fabrication time, high ultimate strain, and unlimited storage life at room temperature. While one of the disadvantages is its difficulty in sufficiently wetting of fibers, as low viscosity cannot be acquired without the use of pressure and heat. Additionally, thermoplastics are more vulnerable to creep, as they have a very low heat deflection temperature compared to thermosetting matrix [12].

The disadvantage of requiring very high temperature processing and costly equipment like an autoclave and a hot press necessitates the need for such a thermoplastic resin that can be processed at room temperature without compromising its mechanical properties [21]. Recently, Arkema, France, developed Elium®, an acrylic thermoplastic resin, which cures at room temperature [22] and can be processed by Vacuum Assisted Resin Infusion (VARI) and Resin Transfer Molding (RTM), typically used for processing low viscosity resins [21, 23]. The composites manufactured with Elium® resin provide several benefits, as they are recyclable, thermoformable, and thermoweldable, substituting the need for epoxies for adhesion [24].

#### 1.2.2.2 Thermosetting Matrices

Polymers where molecules are chemically bound together by covalent bonds or cross-links forming a rigid three-dimensional network structure are known as thermosetting matrices. Due to the presence of permanent cross-linking between the polymer molecules, the thermosetting matrix cannot be melted and reformed using pressure and heat like the thermoplastic matrix. One of the most important advantages of the thermosetting matrices is that they are in the form of a low viscosity liquid, allowing good wetting of the reinforcing fibers in the manufacturing of FRP composites. In addition, other advantages include their high resistivity to creep and stress relaxation, good thermal stability, and high chemical resistance. The disadvantages include lower ultimate strain, long fabrication time during polymerization, and limited storage life at room temperature [12]. Thermosetting polymers/resins normally are of polyester, epoxy, and vinyl ester types.

#### 1.2.2.3 Polyester Resin

Polyester resins provide low cost, ease of handling, and excellent mechanical, electrical, as well as chemical properties which can be modified during the manufacturing process. Depending upon the building blocks of their structure, the polyester resin can be categorized as ortho-polyester and iso-polyester. Based on their end-use application, polyesters can further be divided as general-purpose and specialty polyesters. Ortho-polyesters containing orthophthalic acids are the general-purpose resins with the least cost and are used widely for several applications. However, their strength, moisture, and fatigue properties are slightly lower than those of iso-polyesters. Iso-polyesters, which contain isophthalic acids, are specialty polyesters that are tailored chemically for special material properties such as corrosion resistance, moisture resistance, and fire-retardant characteristics [4]. Iso-polyesters have better structural and environmental characteristics than ortho-polyesters and are also less expensive than vinyl esters [25].

#### 1.2.2.4 Epoxy Resin

Epoxy resins have better heat resistance, less shrinkage during curing, less moisture absorption, and are stiffer, stronger, and tougher compared to polyesters. However, they are more expensive than polyesters and have weaker resistance to UV radiation. They can be used with all types of reinforcing fibers (glass, carbon, aramid, basalt), and by tailoring the manufacturing process, a variety of composites with properties ranging from high strength to high stiffness can be produced. Epoxy resins are widely used in civil engineering applications owing to their better mechanical properties, corrosive and environmental resistance, electrical properties, and high temperature performance [4]. They are also used to bond the FRP composites, as they have good adhesion to many substrates [11].

#### 1.2.2.5 Vinyl Ester Resin

Vinyl ester resins are the unsaturated esters of epoxy resins, which were developed to combine the better mechanical/chemical properties of epoxy resins with the better handling/faster curing properties of polyester resins. Vinyl esters have better mechanical toughness, reduced water absorption and shrinkage, as well as better corrosion and environmental resistance relative to polyesters. Vinyl ester resins are used for extreme environmental conditions because they are more expensive than polyester resins [25]. Vinyl esters are favored in place of polyesters in chemically corrosive environments and where a high degree of moisture resistance is desired, for example, in applications such as exterior structural laminates [4].

## 1.3 FABRICATION/MANUFACTURING OF FRP COMPOSITES

Reinforcing fibers are available in the form of strands/tows, yarns, and roving. Strands are defined as a bundle of continuous glass or aramid fibers, whereas tows are the collection of carbon fibers. Yarns are a collection of twisted continuous strands, whereas roving refers to almost parallel bundles of untwisted continuous

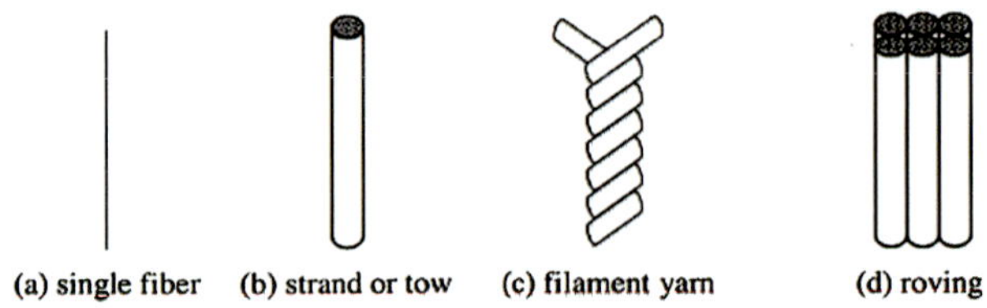


**FIGURE 1.2** Typical forms of reinforcing fibers [4]

strands [4, 11, 26]. These typical forms of reinforcing fibers are shown in Figure 1.2. Apart from being available as continuous fibers (long fibers), reinforcing fibers are also available in the form of discontinuous fibers (short or chopped fibers).

The reinforcing fibers can be either used directly or they can be first transformed into textile or fabric products such as nonwoven fabrics, unidirectional (UD) fabrics, and woven fabrics for the manufacturing of FRP composites. The nonwoven fabrics, available as mats, are made from randomly oriented short or chopped strands of fibers, while the unidirectional (UD) fabrics are made from parallel continuous fibers that are oriented along the direction of the fiber axis. On the other hand, woven fabrics are composed of oriented reinforcing fibers that could be available as biaxial, triaxial, or even quadraxial fabrics. The different forms of available fiber fabrics are shown in Figure 1.3. Finally, FRP composites can be manufactured by introducing

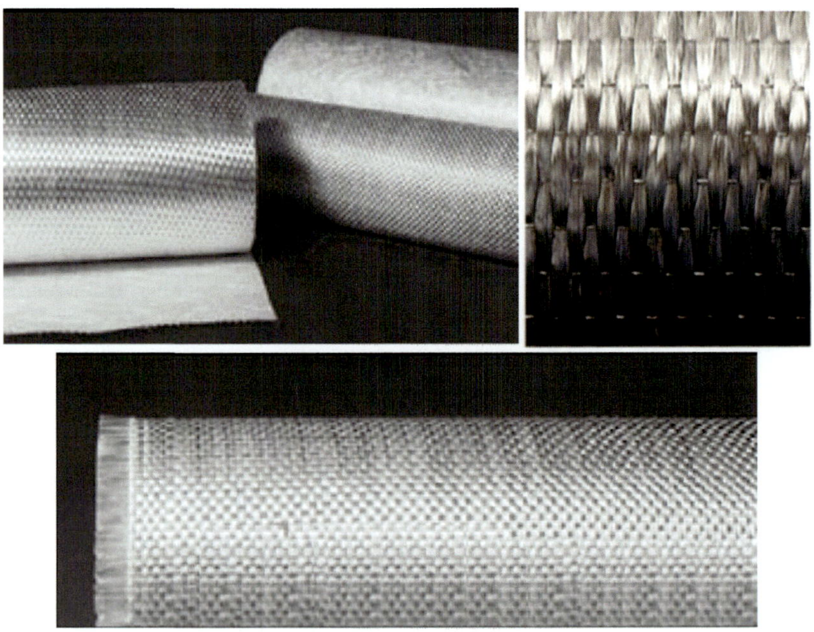

**FIGURE 1.3** Available forms of fiber fabrics [27, 28]

resin into the different forms of available reinforcing fibers or fabrics through several methods discussed in later sections describing the manufacturing methods of FRP composites used in civil engineering applications. In general, the manufacture of FRP composites involves building up layers of resin-impregnated fibers and fabrics via successive wetting and curing process [27].

### 1.3.1 FRP Products Used in Civil Engineering Applications

FRP composites for civil engineering applications are fabricated into a variety of forms such as laminates (wraps/laminates, strips, and plates), bars (rods), cables (strands), grids/gratings, structural shapes, and decks. Inspection of some of these forms and applications such as prestressed FRP concrete structures, FRP structural shapes, and FRP grids used for slabs on the ground (i.e., nonstructural applications) are not covered in this book.

#### 1.3.1.1 FRP Laminates

FRP laminates are used in externally bonded FRP systems for the strengthening of existing structures. These composites can be categorized as wet layup, prepreg, and precured systems. A wet layup system can be differentiated from a prepreg system depending upon where or when the dry unidirectional/multidirectional laminates or fabrics are impregnated and saturated with resin. For a wet layup system, the impregnation takes place on-site, whereas for a prepreg system, the impregnation (or pre-impregnation) takes place at the manufacturer's facility, and the resin matrix is only partially cured. Precured FRP laminates manufactured off-site are available in various forms such as unidirectional strips or plates, multidirectional grids, and shell segments [26]. Figure 1.4 shows an example of externally bonded FRP laminates installed as wet layup and precured FRP plate.

#### 1.3.1.2 FRP Rebars

FRP rebars are embedded into concrete elements as structural reinforcements for new construction. These FRP rebars can be mainly classified into three groups based on the methods used to establish a bond between the rebars and the concrete by creating variations in the geometry of their cross-section and surface [30], as described in the following sections.

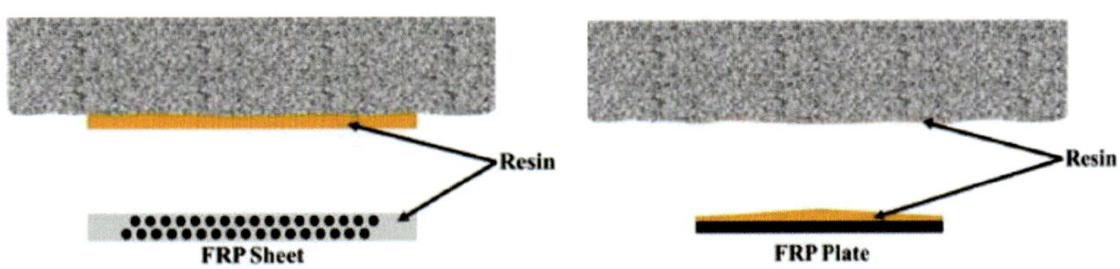


**FIGURE 1.4** Different forms of FRP laminates [29]

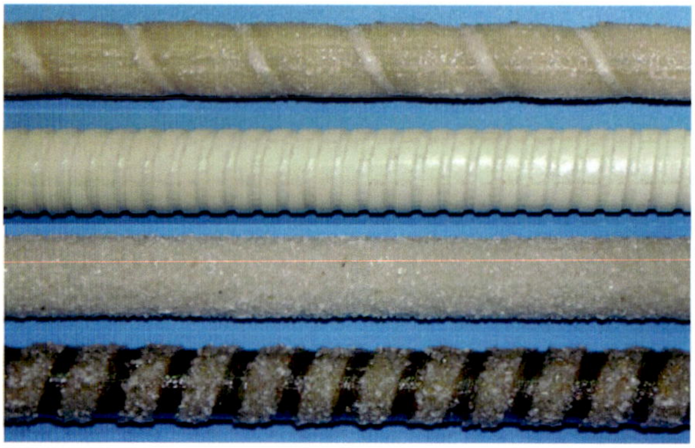

**FIGURE 1.5** Various surface treatments of FRP bars [31]

*Bars with a Round Cross-Section and a Roughened Surface*

The first type of available FRP rebars is designed to increase the adhesion and/or friction between the bar and concrete by roughening the outer circumferential surface of the bar while keeping its cross-section approximately round. An example of this type is shown in Figure 1.5 (third picture from top), where the outer surface of the FRP rebar is roughened by gluing a coating of inorganic particles using an adhesive layer of polymeric resin. Apart from this, there are several other methods available to achieve a roughened surface on FRP rebars with round cross-section which are described in a study by Portnov et al. (2013) [30].

*Bars with a Round Cross-Section and Ribs or Grooves*

Similar to deformed steel rebars, FRP rebars are also available with an undulating outer surface for increasing the mechanical interlocking between the rebar and concrete. Figure 1.5 (second picture from top) shows an example of an FRP rebar with grooves that are formed simply by machining grooves in a smooth FRP rebar. Figure 1.5 (first picture from top) shows an example of an FRP rebar with ribs that are formed by bonding spirally wound FRP strips around a smooth FRP rebar. Similarly, there are several other methods available to achieve an undulated surface on FRP rebars with a round cross-section [30].

*Bars with Noncircular Cross-Sections of Different Shapes*

The third way of increasing the mechanical interlock between the rebar and the concrete is by making the cross-section of the bars noncircular. Noncircular bars could be manufactured; however, they are not used in structural applications. Dowels made with elliptical FRP bars have been used in pavement joints.

In near-surface mounted (NSM) applications, rectangular sections with a large width-to-thickness ratio (i.e., narrow strips) have been used when the width of the cut must be minimized (Figure 1.6). This application is discussed in more detail later in the document.

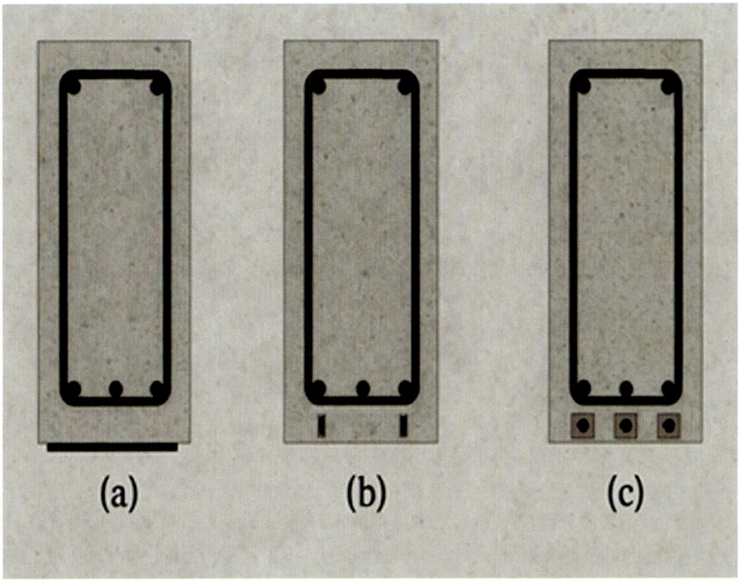


**FIGURE 1.6** NSM application (a) FRP laminate; (b) FRP strips; and (c) FRP bars

### 1.3.2 Manufacturing Methods of FRP Composites Used in Civil Engineering Applications

#### 1.3.2.1 Hand Lay-Up Method

Hand lay-up process, also referred to as wet layup process, as shown in Figure 1.7, is the most commonly used method for fabricating as well as applying FRP composite laminates to other surfaces. The general steps involved in a hand lay-up process are given as follows:

1. Prepare mold/substrate: Remove any dust and dirt from the mold. At times, apply a primer.
2. Apply resin: Apply resin on the mold using a brush.

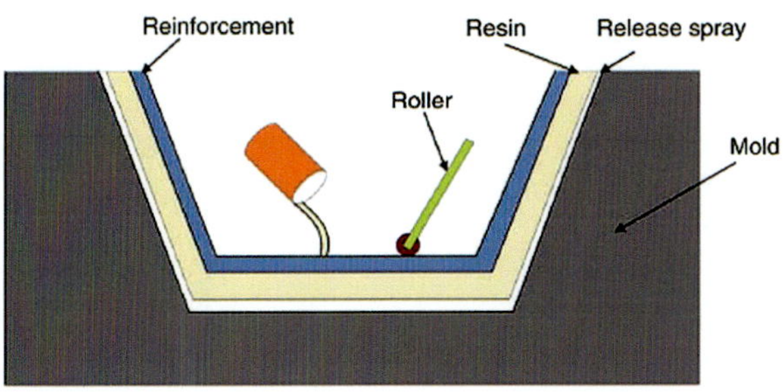


**FIGURE 1.7** Hand lay-up method [32]

3. Lay-up reinforcing fibers: Spread the fibers/FRP laminates over the applied layer of resin.
4. Reapply resin: Apply the resin layer over the fibers/laminates layer using a brush.
5. Consolidation: Apply pressure onto the fiber reinforcements using rollers to ensure good consolidation through uniform distribution of resin and sufficient compaction.
6. Repeat: Repeat steps 3 through 5 until the desired thickness is achieved.
7. Cure: Leave the composite to cure thoroughly, which takes from a few hours to overnight.

Surface preparation, which removes contamination and weak surface layers, is one of the most important steps in adhesive bonding of composite laminates to the concrete elements [33–35]. Improper surface preparation can lead to premature failure of bonded FRP laminates due to rupture/debonding [36]. The application of an adhesive-compatible primer layer after the surface preparation may be desirable on surfaces such as concrete that are porous [33]. Primers are low viscosity materials and aid adhesion by either partially penetrating the pores of a porous surface or by forming a chemical bond between the surface and adhesive or resin that is relatively higher in viscosity [33].

The hand lay-up method requires significantly less capital investment and low-to-moderate labor skills. However, it is very time-consuming with limited control over the quality of work, which can show itself through the presence of voids and porosity, inaccurate thickness, non-uniform fiber-to-resin ratio, and variability in material properties in composites manufactured using this method.

#### 1.3.2.2 Pultrusion Method

Pultrusion, derived from the words "pull" and "extrusion", is the process of manufacturing continuous lengths of FRP composites (laminates, bars, structural shapes, cables) by pulling continuous reinforcing fibers or fiber fabrics through a resin bath and curing them into desired shapes of constant cross section. The schematic diagram of the steps involved in a pultrusion process is shown in Figure 1.8. In the figure, it can be seen that continuous fibers are guided toward a resin bath to first impregnate them with the resin matrix ("wet-out" step). Then they are fed through a preform die to remove excess resin and air and to partially shape them. In the next step, the saturated fibers are pulled through a heated die where the gelation or hardening of the composite is initiated by the heat. The shape of the FRP composite being manufactured corresponds to the shape of the dies, and virtually all types of structural shapes can be produced simply by adjusting the shape of the die [37]. Finally, after the curing is complete, the FRP composites are cut into the desired length.

Pultrusion is the method used in manufacturing FRP rebars. Because of the automation in the manufacturing process, pultrusion is a fast and economical method for mass production of FRP composites. It further provides well-controlled and consistent quality in FRP composite production with easy control of resin content and

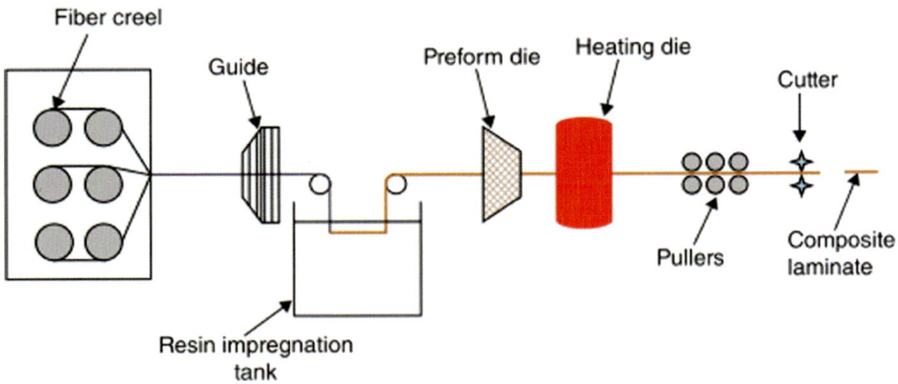


**FIGURE 1.8** Pultrusion method [32]

high fiber volume fractions [27]. However, its main limitation is that the pultrusion process can only manufacture straight composites with a uniform cross section, i.e., the manufacture of components with non-uniform cross sections and curved geometries is not possible [32, 37].

#### 1.3.2.3 Other Manufacturing Methods

There are several other methods, such as resin transfer molding (RTM), vacuum-assisted resin transfer molding (VARTM), and compression molding, that are available to manufacture FRP composites for civil engineering applications. RTM is the process of injecting low viscosity resin under high pressure and temperature, whereas VARTM uses vacuum to impregnate the fiber reinforcements. The latter is a relatively cheaper method. Similarly, compression molding involves preheating the charge and pressurizing it in a mold to form FRP composites.

## 1.4 ADVANTAGES AND LIMITATIONS OF FRP COMPOSITES

Over the years, the use of FRP composites has been on the rise due to the advantages provided by this material. These advantages include high strength, low density, the ability to customize the FRP properties, and good long-term durability. FRP composites have significantly higher strength compared to carbon steel.

FRP composites have less density compared to steel and concrete—about one-fourth the density of steel and about two-thirds that of concrete. Accordingly, their application requires less scaffolding and heavy lifting equipment and hence can be handled by a smaller workforce, and as such provides higher safety. Furthermore, their use as an alternative to steel reinforcements in concrete elements reduces the overall weight of the structure, providing additional advantages.

The properties of FRP can be customized by changing the direction and quantity of fibers. Fabricators and engineers have flexibility in the choice of materials. Various types of fiber and several arrangements (individual fibers, chopped strand mat, woven or stitched fabrics of various geometries) can be used for FRP composites.

This provides fabricators with the ability to use the FRP materials efficiently and offers composites tailored to the customers' needs [38–41].

Much research has been conducted to investigate the durability of FRP materials in long-term performance. The research shows that FRP materials have low degradation in the long term [42] and are good at fatigue resistance. Also, FRP composites have better resistance to salt, water, and other chemicals compared to conventional materials. Oil as well as other heavy hydrocarbons cannot affect FRP composites as much as they can affect conventional materials [43–46].

Although FRP has provided many advantages, there are some downsides to its use. Firstly, the cost of FRP is generally higher than conventional materials. Secondly, FRP has a low coefficient of thermal expansion (and even negative in the case of CFRP), which can become a problem when combined with conventional materials such as concrete. Resin properties are temperature-dependent, and resins are flammable. Low temperatures lead to an increase in strength and brittleness. High temperatures, approaching the glass transition temperature of a resin, cause rapid degradation. This issue can be prevented by choosing a resin type whose glass transition is higher than the anticipated maximum operating temperature [38].

The proprietary aspect of FRP material may also be viewed as a limitation. Although cost and lack of bendability have been considered major issues, the former has significantly changed in recent years.

## 1.5 APPLICATION OF FRP COMPOSITES IN CIVIL ENGINEERING

FRP composites have been used extensively in civil engineering for structural and nonstructural applications in building interiors and exteriors as well as highway bridges.

FRP composites can be used for every aspect of building interiors, from floors to doors. An early example of FRP application in buildings dates back to 1974, when it was used for the cladding of Mondial House in London. Doors, door frames, and window frames have been fabricated in GFRP for many years. Because they have hard smooth surfaces and are electrically non-conductive, FRP materials are practical for places where a high level of hygiene is needed or where electrical or magnetic machines are operated. Their application has been broadened to techniques for the rehabilitation, strengthening, and repair of structures. FRP wraps for seismic retrofit of columns and for strengthening of walls, beams, and slabs have been implemented extensively in the past decades.

FRP materials have also been widely used in highway infrastructure. In a survey conducted by Kim (2017) [37, 47], the application of FRP and the challenges for its use were investigated from 46 transportation agencies (44 US DOTs and 2 Canadian Provincial Ministries of Transportation). According to this survey, 84.4% of respondents had used FRP composites in their projects. The most frequently used application was for bridge girders (CFRP-prestressed concrete girders) with 68.4%, followed by bridge decks and piers (each 63.2%), and piles (26.3%), as shown in Figure 1.9. The term others (23.7%) in Figure 1.9 refers to the use of FRP for concrete

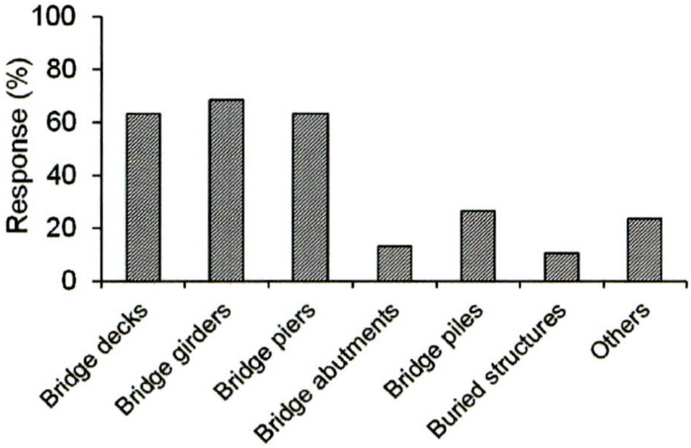


**FIGURE 1.9** Application of FRP in highway bridges (45 answers from 45 responding agencies) [47]

pavement (FRP dowels), a floating bridge, fender piles, bridge drains, shear walls, bent cap wrapping, and culvert invert liners [47].

According to the survey in 2017, GFRP bars were the most typical element utilized as reinforcement for bridge decks and buried structures. Also, GFRP and CFRP laminates were the most common types implemented to strengthen columns and girders.

The survey also demonstrated the application history of FRP composites, as shown in Figure 1.10. A pioneering effort was seen between 1990 and 1995 (21.1%), in which FRP technologies were started to be implemented in civil engineering. A significant increase was achieved from 1996 until 2005 (up to 66%). For the last ten years, 13.2% of the responding agencies started implementing FRP for their construction projects [47].

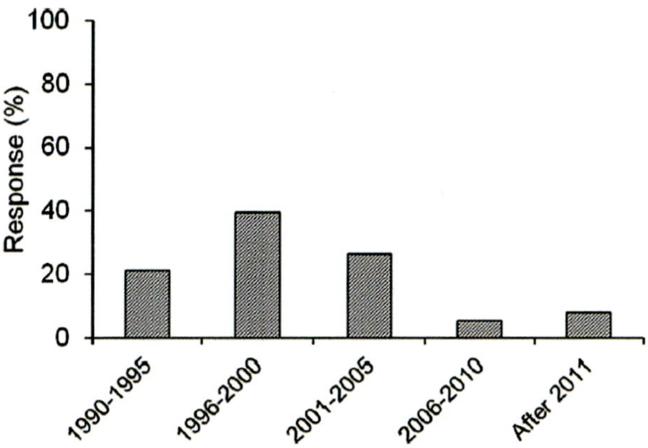


**FIGURE 1.10** First use of FRP composites by various agencies [47]

FRP composite bridge decks where the entire deck structure is made of FRP material, normally in the form of a sandwich panel, have had great applications in the construction of new bridges and replacement of existing decks [25]. There have been other applications where FRP composites have been used alone to function as structural members; however, in the following, as the focus of this project is, the use of FRP materials in concrete structures, i.e., FRP as internal reinforcement or external strengthening for concrete members, is discussed.

## 1.6 APPLICATION OF FRP MATERIALS IN CONCRETE STRUCTURES

There has been significant growth in the use of FRP composites in concrete structures over the last few decades. They have been used for new construction or rehabilitation of existing structures. Generally, FRP applications in concrete structures can be defined into two main categories: (1) FRP bars, rods, and strands as internal reinforcement and (2) FRP laminates, wraps, and near-surface mounted (NSM) FRP bars as external reinforcement.

### 1.6.1 Internal Application (Concrete Structures Reinforced with FRPs)

Over the past two decades, laboratory tests have indicated that FRP bars, strands, and tendons can be implemented practically and successfully as internal reinforcement in concrete members. The industry/university cooperative research paved the way in transferring the know-how of the application of internal FRP reinforcement for concrete from the laboratory to the field. To date, FRP reinforcement has gained good acceptance as internal reinforcement in concrete structures. FRP rebars have a different mechanical behavior compared to steel rebars. FRP materials are linear, anisotropic, and elastic until rupture. They are categorized by high tensile strength only in the fiber direction, with no noticeable yielding [48–50]. Accordingly, concrete members reinforced with FRP bars under flexural loading typically experience large cracking and deflections before failure, which are usually sudden with little warning. The shear strength and the bond performance of the FRP rebars are affected by their anisotropic behavior. Moreover, the performance of the FRP bars in compression is not as good as that in tension. This is due to their nonhomogeneous and anisotropic nature, resulting in a lower compressive modulus compared to that in tension [51]. There is still little experience in using FRP reinforcement in compression members (columns) and for moment frames or zones where moment redistribution is required [52, 53]. However, there has been significant interest in the implementation of CFRP strands in precast prestressed piles to the point that this can be considered a state-of-the-practice. In fact, the Florida Department of Transportation has implemented this technology in numerous field projects and codified its use in Volume 4 of the "Structures Manual 2025 Edition" [54]. Standard Plans for prestressed concrete sheet piles, and square and round bearing piles with CFRP strands are available and can be used without prior FDOT approval. As an example, Table 3.5.1-1 from the "Structural Manual 2025 Edition" is given in Figure 1.11.

Perhaps, more relevant to this book, it can be mentioned that there is an interest in manufacturing and utilizing precast concrete piles with GFRP bars with and without prestressing. In the former case, only partial prestressing is used [55]. The latter case

**Table 3.5.1-1 Concrete Pile Size and Material Requirements**

| Pile Location | Minimum Square Pile Size (inches) | | Minimum Cylinder Pile Diameter (inches) | Material Properties for All Pile Sizes[1] | | |
|---|---|---|---|---|---|---|
| | Vehicular Bridges | Pedestrian Bridges & Fishing Piers | | Strand Type | Spiral Type | Reinforcing Bar Type |
| On land or in water in environments that are Extremely Aggressive due to chlorides[2] | 24[3] | 18 | 54 | Carbon steel, Spec 933 | Carbon steel, Spec 933 | Carbon steel, Spec 933 |
| | 18 | 14 | 54 | CFRP, Spec 933 | CFRP, Spec 932 | FRP, Spec 932 |
| | | | | Stainless steel, Spec 933 | Stainless steel, Spec 931 | Stainless steel, Spec 931 |
| On land or in water in all other environments | 18 | 14 | 54 | Carbon steel, Spec 933 | Carbon steel, Spec 931 | Carbon steel, Spec 931 |

1. See *SDG* 1.4.3 for concrete class and admixture requirements.
2. The use of CFRP or stainless steel strand and reinforcing is required for piles of pile bents located in the water, and may be used at other locations upon approval by the District Structures Maintenance Engineer. If approved by the District Structures Maintenance Engineer, piles of pile bents with carbon steel strand and reinforcing (of the minimum sizes shown with admixtures) may be allowed in the water for minor widenings. This decision is dependent upon site-specific conditions, anticipated structure life, and the history of piles in the vicinity.
3. If approved by the District Structures Maintenance Engineer, a minimum pile size of 18 inches may be allowed for piles of pile bents for minor widenings in the water. This decision is dependent upon site-specific conditions, anticipated structure life, and the history of piles in the vicinity.

**FIGURE 1.11** Concrete pile size and material requirements for pecast prestressed CFRP piles based on Structural Manual 2025 Edition [54]

(i.e., no prestressing) could represent an effective and economical solution for coastal construction [56].

### 1.6.1.1 FRP Bars, Strands, and Grids

FRP rebars are appropriate alternatives to steel, stainless steel, and epoxy-coated steel bars in reinforced concrete applications when durability, electromagnetic transparency, or ease of demolition in temporary applications is essential for the project. FRP bars, strands, and, more recently, meshes (Figure 1.12) have been successfully implemented as structural reinforcement for concrete members in bridge and building projects in the past three decades [57].

The majority of applications employ FRP rebars to reduce the possibility of corrosion in concrete members that operate in aggressive environments or areas that are exposed to deicing salts. The service life for these members is dependent on the durability of the internal reinforcement. Although their initial cost, which includes their raw materials and manufacturing cost, might be slightly more than that of conventional steel, their application in concrete structures subjected to a harsh environment can improve the service life of the structure and lower the overall life cycle cost. In the following Table 1.4 [52], examples of the application of FRP materials as internal reinforcement in concrete structures are presented.

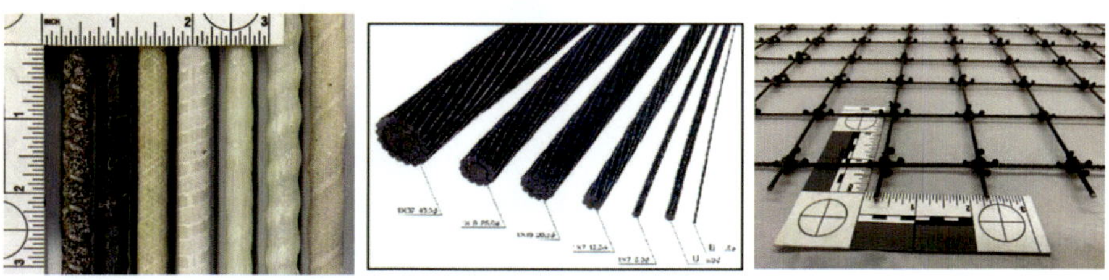

**FIGURE 1.12** FRP materials applicable for internal reinforcement of concrete members

**TABLE 1.4**
**Examples for the application of FRP materials as internal reinforcement in concrete structures**

**Bridges at sea, port infrastructure, retaining/sea walls, and dry docks**

| | |
|---|---|
| Concrete bridge reinforced with CFRP grid (Fukushima Prefecture, Japan) |  (b) |
| Application of FRP in Honoapiilani highway retaining sea wall south (Lahaina, Maui, Hawaii) |  |
| Application of FRP dowel bars in concrete pavements (Port of Rotterdam, The Netherlands) |  |
| Application of GFRP bars for Pearl Harbor dry docks (Honolulu, Hawaii) |  |

*(Continued)*

## TABLE 1.4 (CONTINUED)
## Examples for the application of FRP materials as internal reinforcement in concrete structure

Application of GFRP bars for a secant-pile seawall (east of State Road A1A (SR A1A) at Flagler/Beverly Beach in the northeastern section of Flagler County, FL)

The use of FRP reinforcing bars for piers and bent caps (Broward County, FL, USA) [57]

**Bridge decks and railings where deicing salts are used**

Application of GFRP bars in concrete bridge deck (Morristown, Vermont)

*(Continued)*

## TABLE 1.4 (CONTINUED)

**Examples for the application of FRP materials as internal reinforcement in concrete structure**

Application of GFRP bars in concrete bridge deck (Cookshire-Eaton, Quebec, Canada)

GFRP cages prior to casting a bridge railing (Greene County, Missouri)

**FRP reinforcement applications when equipment sensitive to electromagnetic fields exists [52]**

Application of GFRP bars in concrete slab for MRI rooms in hospital (York, Maine)

*(Continued)*

## TABLE 1.4 (CONTINUED)

**Examples for the application of FRP materials as internal reinforcement in concrete structure**

| | |
|---|---|
| Application of GFRP bars in concrete slab (Oran, Algeria) |  |
| GFRP reinforced-concrete rail track structure (Miami, Florida) |  |
| **FRP rebars applications for reinforcing the temporary concrete structure [52]** | |
| GFRP reinforcement cage for soft-eye construction at a manufacturing plant (Angri, Italy) |  |
| GFRP reinforced-concrete soft-eye for Washington Dulles International Airport people mover (Dulles, Virginia) |  |
| GFRP reinforced-concrete soft-eye for Beacon Hill light rail transit (Seattle, Washington) |  |

(*Continued*)

**TABLE 1.4 (CONTINUED)**

**Examples for the application of FRP materials as internal reinforcement in concrete structure**

| | |
|---|---|
| GFRP soft-eyes for tunnel excavation (a) and (b) view of bars when TBM emerges (London, UK) |  (a) (b) |
| **Other applications [57]** | |
| GFRP reinforced concrete for Ice Harbor Lock and dam fish weir (Walla Walla, Washington) |  |
| Construction of a residential home in Great Harbour, Berry Island, Bahamas, with GFRP reinforcement used for all structural concrete elements including the masonry walls |  |
| BFRP meshes as secondary reinforcement in slabs-on-ground (Marathon, FL, USA) |  |

### 1.6.1.2 FRP Tendons for Prestressing and Posttensioning Applications

FRP tendons can be fabricated in the form of rounded or flat bars, strands, ropes, or cables. Various surface textures are feasible, which include smooth, braided, dimpled, rough, sand-coated, indented, etc. The most common types of FRP tendons are Aramid FRP (AFRP), Carbon FRP (CFRP), and Glass FRP (GFRP) [58].

Field applications of bridges with FRP tendons have been consistently increasing since the mid-1980s. The first GFRP prestressed bridge was Lunensche Gasse Bridge, built in Dusseldorf, Germany, in 1980. The first bridge prestressed with CFRP tendons was constructed in Japan in 1988. In the United States, the first bridge that used CFRP and GFRP cables was built in Rapid City, South Dakota, in 1992. Table 1.5 indicates some of the early bridge projects before the year 2000 in the world that implemented FRP tendons. Figure 1.13 shows an example of the application of FRP post-tensioning tendons/cables for a bridge located in the City of Southfield, US [58].

### 1.6.2 External Application (Rehabilitation)

The use of FRP materials as an external application has been mostly for the rehabilitation of structures, which may be necessitated in the following situations:

1- There are deficiencies at the design stage. These include design errors, use of poor-quality materials, and insufficient safety factors;
2- The load carrying capacity in an in-service element is changed. This can occur when an increase in the safety requirement is needed (upgrading of structural design standards) or when the applied load is increased due to the of the structure or increased live load;
3- Degradation and aging of the materials have occurred which requires increasing the bridge rating factor not to be posted; and
4- There is a need for strengthening after extraordinary events including seismic or fire events [61]

Rehabilitation of the concrete structures can be categorized into strengthening, retrofitting, and repairing the concrete members. The term repairing is used when the FRP materials are implemented to address the poor condition of the structural elements. Strengthening is specific to the cases in which the application or addition of FRP would enhance the performance level of the existing designed elements. The term retrofit is used when FRP materials are utilized for improving the performance of an element for seismic and other high-risk hazards such as hurricanes [62].

#### 1.6.2.1 Strengthening, Retrofitting, and Repair

External FRP systems are generally bonded to the tension side of the concrete girders, beams, and slabs to enhance the flexural strength. They can also be used to provide additional shear strength when applied on the sides of beams and girders. In seismic zones, FRP wraps can be used for columns to increase ductility due to the induced confinement of the concrete [63]. In the following, various methods practiced for external application of the FRP materials for concrete elements are discussed:

#### 1.6.2.2 FRP Laminate

Continuous FRP laminates weighing 200-410 g/m$^2$ have been employed over the years for the repair and rehabilitation of deteriorated concrete structures (JCI 1997;

**TABLE 1.5**

**Examples of FRP pre-tensioned and post-tensioned concrete structures before the year 2000 [58, 59]**

| Tendon | Name | Description | Dimensions | Country | Date |
|---|---|---|---|---|---|
| CFRP | Shinmiya Bridge | Pre-tensioned slab bridge | Span 5.7 m 6 tendons at bottom. 2 at top of slab | Japan | 1988 |
| | Ludwigshafen Bridge | 4 spans, post-tensioned. single-span highway bridge | Total span 85 m 2 straight and 2 curved spans, R= 62.8 | Germany | 1991 |
| | Rapid City Bridge | Precast. post-tensioned slab bridge. at a cement plant | Span 9 m. width 5.2 m 180 mm thick slab, 2/3 FRP prestressed | USA | 1992 |
| | Beddington Trail Bridge | Precast, pre-tensioned, two-span highway bridge | 22.8 and 19.2 m 6 of 26 bulb-T girders with FRP | Canada | 1993 |
| | Headingley Bridge | Precast, pre-tensioned girders | 5 spans, each 32.5m 4 of 40 AASHTO I-girders with FRP | Canada | 1996 |
| GFRP | Lunensche-Gasse bridge | Single-span slab bridge | Span 6.55 m 100 Polystall GFRP. unbonded. tensioned | Germany | 1980 |
| | Ulenberg-Strass Bridge | Two-span highway bridge | 59 cables, each 19–7.5 mm Polystall | Germany | 1986 |
| | Khabarovsk Bridge | Five precast, prestressed girders | Span 15 m, 24 GFRPs | Russia | 1989 |
| | Notsch Bridge | Three-span highway bridge | Spans 13. 18, 13 m slab thickness: 650 mm | Austria | 1991 |
| AFRP | Tabras Golf club Bridge | Three-span. pretensioned slab bridge | Span 11.8 m. 2.4 m wide. 3 of 21 girders, AFRP prestressed | Japan | 1990 |
| | Birdie Bridge | Post-tensioned pedestrian bridge with suspended slab | Span 54.4 m. 2.1 m wide. 16 cables with 8 bundled AFRPs | Japan | 1990 |
| | Sumitomo Bridge | Highway bridge, pre-tensioned slab/post-tensioned box girder | Span of slab 12.5 m. box girder span 25 m. 3 AFRP strands for the slab and 26 AFRP strands for girder | Japan | 1991 |

**FIGURE 1.13** Example of bridge prestressed with CFRP tendons in the United States (City of Southfield's new Bridge Street Bridge) [60]

JSCE 2001; ACI2001). The application of steel plates for the repair and strengthening has been limited due to two main reasons. Firstly, steel plates tend to separate from the concrete interface, and hence might not monolithically behave with the concrete. Secondly, ensuring the proper adhesion between the steel plates and the concrete surface can be an issue due to the uneven nature of the concrete surface. The development of the FRP laminate/jacket system addresses both issues as laminates adhere to the concrete closely [64]. The corrosion resistance provided by FRP material in corrosive environments can also be added as another advantage.

Various steps are needed for the implementation of the FRP laminate/jacket system for concrete members (Figure 1.14):

- Surface preparation: This includes rounding the corners of the members, abrading uneven surfaces with the use of disc sanders, and impregnation of cracks and application of smoothing agents. In some cases, using primer coats is necessary.
- Applying FRP sheets along with impregnating resins and hardeners: The surface of the element is primed with resin or any applicable bonding agent before placing the laminates. The entrapped air is removed in this procedure using rollers while coating with resins. If multiple layers are used, the process should be repeated after placing each layer.
- Curing of FRP sheets.
- The use of finishing coats on the FRP sheets [64].

Various schemes for installing FRP laminates can be practiced (Figure 1.15), which include side bonding, U-wrapping, and full wrapping. It goes without saying that FRP laminates can be more easily wrapped around columns or chimneys where all the sides are accessible compared to the beams where the compression face is usually not easily accessible.

The fibers in laminates can be placed either in one direction or two directions, depending on the design purposes. One-way FRP reinforcement (the fibers are aligned to the element axis) is mostly used for the columns and beams, whereas two-way reinforcement is mainly used for slabs.

**FIGURE 1.14** Application of external FRP [65]

The application of FRP laminate in concrete structures has been generally for flexural, shear, axial, torsional, and seismic rehabilitation of the element, which will be discussed in the following paragraphs.

### *1.6.2.2.1 Flexural applications*

FRP laminates, fabrics, or strips can be bonded to the soffit of reinforced concrete beams to improve their flexural performance. Quite a few research studies have investigated their effectiveness over the years. In 1991, Ritchie et al. [66] tested several RC beams strengthened with CFRP and GFRP laminates. The test specimens were 2.75 m (108.27 inches) long and were tested in bending. Their work indicated that strengthened beams achieved a 40–95% increase in ultimate strength and a 17–95% increase in stiffness compared to the similar beams without strengthening. In another study, Meier et al. [30] experimentally investigated the effectiveness of CFRP laminates by testing 60 small-scale RC beams under a four-point bending setup. The results of their work showed that the flexural capacity of strengthened beams can be increased up to 100% over the beam without strengthening. Arduini and Nanni [67] experimentally studied pre-cracked beams strengthened with several CFRP laminates. They found that fiber direction, fiber stiffness, and the number of

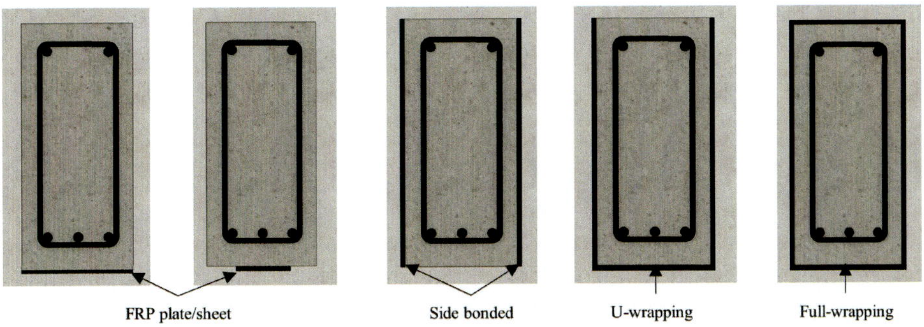


**FIGURE 1.15** Various methods for installing FRP laminate to the concrete element

laminates are among the factors that can affect flexural capacity. H Toutanji, Zhao, and Zhang [68] studied the effect of the number of bonded FRP laminates on the load-bearing capacity of RC beams. Several reinforced concrete beams strengthened with three to six layers of CFRP laminate were tested in their work. They reported up to a 170.2% increase in the load-carrying capacity of the strengthened beam compared to the beam without strengthening. Also, they found that the increase in the number of bonded sheets not only can reduce the ductility of the strengthened beam, but it can also affect the failure mode and change it to a brittle type.

Several failure modes were observed for the members strengthened with FRP laminates during experimental testing. Rupture of FRP laminates, debonding of

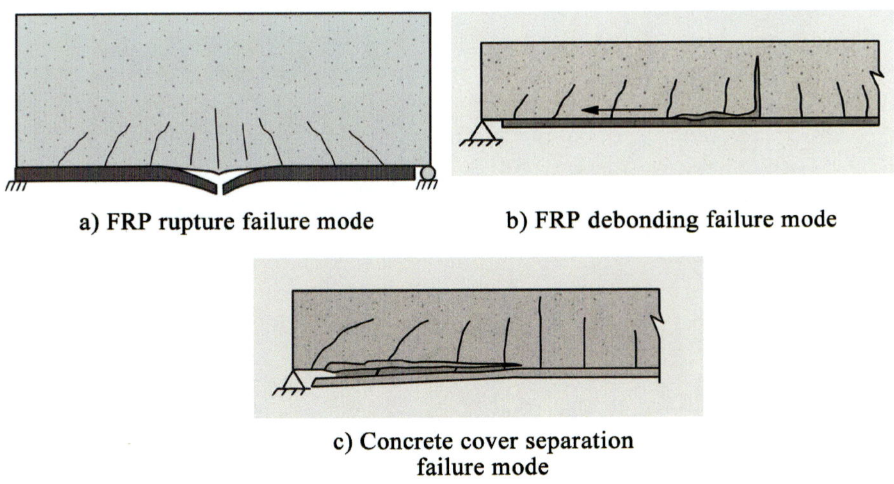


**FIGURE 1.16** Failure modes for RC member strengthened with FRP laminate [69]

FRP laminates from the surface of concrete, and cover delamination (concrete cover separation) are the common failure modes (Figure 1.16) according to ACI 440.2R-17 [5] design guidelines.

Rupture of the FRP laminates will happen when the strain in the farthest FRP laminate reaches its maximum before the concrete reaches its crushing strain. Debonding of FRP laminate or cover separation will occur when the concrete substrate cannot resist the axial force produced in the bending. FRP debonding is often started by flexural-shear or flexural cracks in the maximum moment region. The cracks then propagate through the bonding material (cement matrix and epoxy adhesive) along the length of FRP. These cracks develop and broaden under loading and result in debonding. Concrete cover separation normally starts with a crack close to the end of FRP laminates in which high-stress concentration exists. The cracks then progress along the level of the steel tension reinforcement, which causes the separation of concrete cover from the member [69].

### 1.6.2.2.2 *Shear Application*

RC members can be strengthened against the shear forces by using FRP materials. This includes implementing individual and narrow FRP laminates bonded on the side of RC members acting in parallel with steel stirrups to improve the shear capacity (Figure 1.17). Several studies investigated their effectiveness in improving the

**FIGURE 1.17** FRP laminates for shear strengthening

shear capacity of RC members. Chajes et al. [70] tested under-reinforced concrete beams to find the effectiveness of externally bonded FRP laminate in enhancing the shear capacity. They concluded that shear capacity in strengthened beams could be increased up to 150% in comparison with similar beams without strengthening. El-Maaddawy and Chekfeh [71] and Carolin and Taljsten [72], in their research studies, indicated that CFRP laminates can significantly enhance the shear capacity of RC beams. They also found that implementing proper end anchors can help postpone or prevent CFRP debonding and thus, increase the gain in shear strength.

#### *1.6.2.2.3 Axial application*

FRP laminates have been implemented for axial strengthening of RC elements (Figure 1.18), especially bridge columns and piers [73–75]. Lateral confinement of concrete columns through wrapping FRP laminates can enhance the compressive strength of the member and prevent longitudinal steel buckling. Yamamoto [76] tested several

**FIGURE 1.18** Illustration of the use of FRP laminate for axial strengthening in RC columns

concrete columns wrapped with FRP materials to investigate the effect of FRP on uniaxial concrete strength. He concluded that the strengthened columns could achieve three times the ultimate strength of columns without strengthening. FRP laminates have also been in use for strengthening walls. Antoniades, Salonikios, and Kappos [77] conducted experimental tests on several reinforced concrete cantilever walls strengthened with FRP laminates. The study found that strengthened walls achieved much higher drift capacities compared to the walls without strengthening. Mosallam and Nasr [78] evaluated the effect of CFRP laminates on the structural performance of shear walls with openings. Their work indicated a 20–28% increase in the average strength gain and ductility of the retrofitted walls over the walls without strengthening.

#### *1.6.2.2.4 Torsional Application*

There have been limited research studies regarding the torsional application of FRP composites for RC members. This is because RC elements generally experience shear and flexural loads, and most provisional codes tend to neglect the torsional effect when there is adequate shear reinforcement. Furthermore, the complicated test setup required for performing the torsional tests can make such tests very difficult. Among the limited research studies, Salom, Gergely, and Young [79], in an experimental study, investigated the torsional behavior of several RC spandrel beams wrapped with CFRP composites. Their work revealed that all beams wrapped with CFRP achieved a 50% increase in torsional capacity compared to the similar control beams (without CFRP composites). In 2013, Khalaf and Bayer [80] carried out several tests to investigate the effectiveness of CFRP laminates in the torsional performance of RC beams. Their work included implementing two configurations of CFRP wrapping: fully and partially wrapped laminates. They concluded that beams fully wrapped by CFRP achieved twice the ultimate torque of those similar control beams (without wrapping). They also found that using a proper anchorage system for the U-jacket laminate can improve the overall response significantly. Mohammadizadeh and Fadaei [81] carried out an experimental study on several beams made of high-strength concrete strengthened with CFRP composites to investigate their effectiveness in enhancing the torsional capacity. The results of their work indicated that all strengthened beams achieved an improvement in the torque strength. This improvement depended on the amount of added CFRP reinforcement and their strengthening configuration. They also found that implementing an anchorage system for U-wrapped beams would avoid CFRP delamination, hence could provide the same torsional capacity as a fully wrapped beam with one-layer CFRP laminate.

#### *1.6.2.2.5 Seismic Application*

Although most studies investigated the static and monotonic behavior of RC members strengthened with FRP materials, their seismic response has also been studied. The cyclic response of RC elements strengthened with FRP composites has been thoroughly investigated, as FRP was primarily developed as a retrofitting technique, especially for bridges and buildings damaged during earthquakes. The effectiveness of FRP strengthening for seismic application depends on the type of application, that is, applying FRP to columns, beams, etc. In 2003, Ye et al. investigated the effectiveness of CFRP laminates for enhancing the seismic performance of RC columns. Eight specimens were constructed and tested under cyclic load. Their work showed CFRP wraps could increase the seismic performance of the RC columns due to the provided confinement [82]. Barnes and Mays [83] tested three reinforced concrete beams strengthened with GFRP plates to investigate their cyclic behavior. They found that strengthened beams failed primarily in a flexural mode without any noticeable differences with similar control beams (without strengthening). Papakonstantinou, Petrou, and Harries [84] investigated the effectiveness of implementing GFRP laminates on the fatigue performance of several RC beams. Their work indicated that beams failed mainly because of debonding of the GFRP laminates and the yielding of steel reinforcement. Tanarslan and Altin [85], under cyclic loading, tested several identical deficient RC T-beams strengthened in different configurations, including U-wrap, U-jacketed, L-shaped, side-bonded, and double L-shaped. Their work showed that using three and two layers of CFRP strips could enhance the failure load by approximately 15% and 6%, respectively. Totuanji et al. [86] experimentally investigated the cyclic behavior of concrete beams strengthened with CFRP laminates. They found that both fatigue life and load capacity of RC beams could be enhanced by the addition of CFRP laminates. Hosny et al. [87] tested several T-beams strengthened with a hybrid combination of GFRP and CFRP laminates under cyclic loads. They concluded that strengthened beams reached higher ultimate loads (10–30%) compared to the similar beams without strengthening.

#### 1.6.2.3 Prestressed FRP Laminate

Although FRP laminates have been used widely over the past decades, the research studies have indicated that their effectiveness and reliability may be reduced in long-term performance. FRP laminates do not participate in resisting dead load, and they only participate in bearing live loads. Furthermore, conventional FRP laminates tend to peel off before the concrete element reaches its ultimate state [64]. To address these issues, a method implementing prestressed FRP laminates has been developed [64].

The following steps are needed for the implementation of this method:

- Positioning line, slotting, and surface preparation;
- Placing the prestressing jack and applying tension force to FRP laminate;
- Applying the proper adhesive material along the surface of the FRP laminate and bonding it to the concrete surface;
- Prestressing of the FRP laminate and anchoring it to the concrete surface using several steel strips and bolts;
- Applying protective treatment of anchor devices and FRP laminate against fire and corrosion.

Various research studies investigated the performance of prestressed continuous FRP laminates for RC members. These studies indicated the advantages of this system over the conventional FRP laminate (non-prestressed) for several reasons,

**FIGURE 1.19** NSM FRP system

including the low possibility of debonding in long-term performance and providing higher flexural capacity. In their investigation, Wu et al. (2001) found that strengthened RC members with prestressed FRP laminates provide for higher cracking load, yield load for the reinforcing bars, and breaking loads compared to the similar RC members strengthened with non-prestressed FRP laminates [49].

#### 1.6.2.4 Near Surface Mounted Bar (NSMB)

Over the past decades, near surface mounted (NSM) FRP bar system has attracted an increasing amount of research and practical applications for strengthening reinforced concrete structures. In this technique, grooves are first cut into the concrete surface of an RC member, and FRP bars are then placed into the grooves, then bonded to the concrete with the use of proper adhesive materials (typically epoxy paste or cement grout) (Figure 1.19). What is herein called "NSM bar" was previously given other names such as embedded reinforcement or grouted reinforcement [88, 89]. The first application of NSM steel bars for strengthening RC structures dates back to the late 1940s and was in Europe [88]. NSM techniques have also been used with stainless steel bars for strengthening arch bridges and masonry buildings [90]. The use of FRP bars/strips instead of regular steel bars has provided many advantages, including better corrosion resistance, increasing the speed and ease of installation, and reducing the groove size due to the higher tensile strength of FRP.

Most applications of NSM with FRP bars for concrete elements have been for flexural strengthening [91–95]. De Lorenzis et al. [91] found that this method can increase the load-carrying capacity of RC beams in the range of 25–44%. After long-term tests, Astorga et al. [92] concluded that the NSM system with CFRP bars enhances the ultimate and yield load-bearing capacity of concrete bridge slabs by 41–20%, respectively. Barros et al. [93] investigated RC beams retrofitted with NSM-CFRP. They found that this method was more effective than the externally bonded (EB) CFRP method in the ultimate load-carrying capacity of concrete members. Some applications of FRP NSM for flexural strengthening included prestressing the bars/strips before bonding [96–100].

Another application of the NSM bars using FRP material has been for shear strengthening of existing concrete elements (simply supported RC beams and deep RC T-beams) [101–104]. Barros and Dias [101] experimentally investigated the NSM shear strengthening technique with the use of CFRP strips. They found that this method improves the load-carrying capacity effectively for both reinforced concrete and unreinforced beams. Rizzo and De Lorenzis [102], after experimental tests on simply supported RC beams reinforced with the NSM-FRP system, found that the shear contribution of the system is 2.6 to 2.8 times higher than that of the EB-FRP system.

NSM FRP technique has a number of advantages compared to the externally bonded FRP laminate system: (1) it can be installed faster at the site as it does not require surface preparation other than grooving, (e.g., plaster removal and rounding corners are not necessary); (2) the possibility of debonding and premature failure of NSM FRP technique is much lower; (3) anchoring NSM bar to the concrete is much simpler, which is attractive for the flexural strengthening of beams and columns in rigidly jointed frames where the maximum moments usually occur at the ends of the member; (4) prestressing NSM technique is simpler; (5) it is more durable in the long term due to the fact that NSM FRP bars are protected by the concrete (or other bonding agent) cover, so that they are less exposed to mechanical and environmental damage, accidental impact, and fire; and (6) the aesthetic of the strengthened structure is almost unchanged [105].

#### *1.6.2.4.1 FRP Reinforcement*

Most studies have indicated the application of carbon FRP as NSM reinforcement for strengthening concrete structures. The use of Glass FRP in this technique has been mostly for the strengthening of timber and masonry structures. It should be noted that the tensile strength for the CFRP bar is much higher than a GFRP bar, so that for the same tensile capacity, CFRP bars have a smaller cross-sectional area compared to the GFRP bars. As such, its application in this technique requires a smaller groove size, leading to simpler installation and saving the groove-filling materials. The FRP bars in the NSM technique can be in square, rectangular, round, and oval cross-sectional shapes, as well as strips. Different cross-sectional shapes provide different advantages and accommodate various options for practical applications. For example, a round bar is more readily available in the market than other types and can be more simply anchored in prestressing procedures. Square bars, for a given size of square groove, maximize the bar sectional area. Narrow strips maximize the surface area-to-sectional area ratio for a specific volume and hence reduce the risk of debonding. In practical applications, selection of the shape of bars depends on the depth of the cover, the availability, and the cost of a particular type of FRP bar [5, 105].

#### *1.6.2.4.2 Groove Filler*

The role of the groove filler is to transfer the stress between the concrete and the FRP bar. The most important factors in selecting the proper groove filler are its tensile and shear strength. The first one is especially important when the embedded bars have a deformed surface because high circumferential tensile stresses can be produced in the cover formed by the groove filler due to the bond action. The latter is important when the bond capacity of the NSM reinforcement is governed by the cohesive shear failure of the groove filler. To minimize the possibility of debonding in the NSM FRP

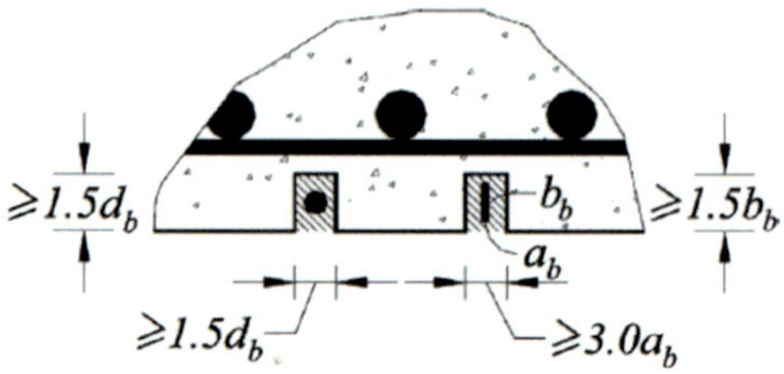


**FIGURE 1.20** Minimum dimensions of grooves [5]

strengthening technique, the proper filler material should be selected. Currently, the most common and best-performing filler material is a two-component epoxy.

#### *1.6.2.4.3 Groove Dimensions*

The groove dimension is to be calculated properly to provide the FRP bar within the groove with the required bond stress. Otherwise, debonding of the FRP bar from the groove filler could be the failure mode. Figure 1.20 shows the minimum groove dimension for the NSM system according to ACI 440.2R-17 [5].

#### *1.6.2.4.4 Groove Position*

If a single NSM FRP bar needs to be used on the tension side of RC members, it should be centrally placed over the beam width. In the case of using two or more NSM bars, the distance between two adjacent NSM bars and the space between the edge of the member and the adjacent bar are important and should be designed suitably [5].

### 1.6.2.5 Connections

As a part of ongoing research at Florida International University (FIU), Khedmatgozar Dolati and Mehrabi (2020) have been investigating the application of FRP laminate and NSM FRP system for splicing precast concrete prismatic elements [106–109].

# 2 Available Standards and Specifications for Design and Installation

## 2.1 STANDARDS AND SPECIFICATIONS

When designing a concrete element using FRP composites, whether as internal reinforcement or as a structural strengthening system, the applicable design specifications vary depending on the intended application and the governing jurisdiction. In North America, the American Association of State Highway and Transportation Officials (AASHTO) and the American Concrete Institute (ACI) provide widely adopted documents—AASHTO for transportation infrastructure and ACI for buildings and other structures. Canadian practices follow the Canadian Standards Association (CSA), which offers comprehensive guidance on various FRP systems. In Asia, while several countries are developing national guidelines, the most internationally recognized standards are from the Japan Prestressed Concrete Institute (JPCI). The United Arab Emirates, for example, does not have its own FRP codes but adopts ACI and AASHTO provisions through regional design manuals. European and international efforts, led by the European Committee for Standardization (CEN), the International Federation for Structural Concrete (fib), and the International Organization for Standardization (ISO), provide harmonized approaches to design, durability, and material performance.

The following sections are organized by application: 2.1.1 Internal Reinforcement, 2.1.2 External Strengthening, and 2.1.3 Material Specifications. Each includes North American and international standards relevant to serviceability and strength limit states.

### 2.1.1 Internal Application

#### 2.1.1.1 North America

The internal use of FRP reinforcement, specifically GFRP bars, is governed by the AASHTO LRFD Bridge Design Guide Specifications for GFRP-Reinforced Concrete (second ed., 2018). This document provides flexural design procedures, strength reduction factors ($\varphi = 0.55$ to $0.75$ depending on failure mode), and serviceability criteria. FRP bars are not allowed to contribute to compressive strength, and the specification emphasizes the brittle-elastic behavior of FRP, which necessitates serviceability control through crack width and deflection limits.

DOI: 10.1201/9781003646969-2

The GFRP bars must comply with ASTM D7957, which specifies mechanical properties, geometric tolerances, and durability requirements. The AASHTO guide specifies environmental reduction factors (typically 0.7 to 0.8) and mandates creep rupture limits by setting the maximum allowable sustained stress at 30% of the guaranteed tensile strength. Minimum concrete cover for fire resistance is required (1.5–2.0 times bar diameter), but no specific fire design methodology is offered.

ACI 440.11-22 was the first U.S. building code dedicated specifically to the design of concrete structures reinforced with glass fiber-reinforced polymer (GFRP) bars. It applies to non-prestressed, cast-in-place, and precast concrete elements including beams, slabs, columns, walls, and foundations using solid, round GFRP bars conforming to ASTM D7957. The code is built on the recognition that GFRP is a linear-elastic, brittle material, unlike steel, and therefore emphasizes strain-based design and serviceability rather than ductility. Strength reduction factors (φ) range from 0.55 to 0.75 depending on the failure mode, and FRP bars are not permitted to contribute to compression resistance. Due to the lower modulus of elasticity of GFRP, deflection and crack width control are essential, with specific serviceability limits imposed to ensure long-term durability and aesthetic performance. Environmental reduction factors are applied to account for durability concerns, and sustained stress is limited to 30% of the guaranteed tensile strength to prevent creep rupture. In terms of fire performance, minimum concrete cover is prescribed (1.5 to 2.0 times the bar diameter), although no explicit fire design methodology is provided. The code also limits the use of GFRP in seismic applications: it is permitted in Seismic Design Category (SDC) A for all elements, conditionally allowed in SDC B and C if not part of the seismic force-resisting system, and prohibited in SDC D through F. While it shares some principles with the AASHTO LRFD Guide Specifications for GFRP-Reinforced Concrete (2018), ACI 440.11-22 is specifically tailored for buildings and offers broader provisions for vertical and lateral load-resisting members within non-seismic structures.

#### 2.1.1.2 Canada

The CSA S806-12 (R2017) serves as Canada's primary document for design and construction using fiber-reinforced polymers. It covers both internal GFRP reinforcement and externally bonded systems. Compared to AASHTO, CSA allows broader use of FRP types (GFRP, CFRP, BFRP, AFRP) and explicitly defines fire, fatigue, and environmental factors. Its strength reduction factors and environmental durability considerations are aligned with ISO recommendations.

#### 2.1.1.3 Europe

The CEN/TS 19101:2022, a European Technical Specification, presents a unified approach for the design of concrete members reinforced with FRP. It covers GFRP, CFRP, and BFRP, and adapts Eurocode-based limit state design philosophy, including partial safety factors and durability considerations. Flexural and shear design models are based on strain compatibility, and detailed provisions are made for crack control, deflection limits, anchorage behavior, and environmental exposure classes.

The fib Bulletin 40 complements CEN/TS 19101 by providing comprehensive background on FRP-RC design, including experimental validation, long-term performance, and fire testing. It discusses fire degradation, loss of bond strength, and residual stiffness at elevated temperatures.

#### 2.1.1.4 Asia

The JPCI guidelines [110] also include internal FRP reinforcement provisions, particularly for precast prestressed bridge elements. The design models consider pre-tensioned CFRP tendons and provide stress limits for transfer and service conditions. Serviceability is controlled via allowable crack width and deflection limits, consistent with AASHTO and CSA approaches.

### 2.1.2 External Application

#### 2.1.2.1 North America

The design of externally bonded fiber-reinforced polymer (FRP) systems for the rehabilitation and strengthening of highway bridge structures is primarily governed by the Guide Specifications for Design of Bonded FRP Systems for Repair and Strengthening of Concrete Bridge Elements (AASHTO, second ed., 2023). This specification includes provisions for both wet layup systems and pre-cured laminates bonded to the surface of reinforced concrete (RC) or prestressed concrete (PC) elements. It also permits the use of near surface mounted (NSM) FRP systems. A notable update in the 2023 edition is the introduction of an environmental reduction factor (fEM), which accounts for long-term environmental effects on FRP materials. This factor varies between 0.65 and 0.95 depending on environmental exposure categories. In addition to this, the revised guide enhances alignment with ACI 440 strain-based design approaches, includes specific provisions for flanged sections in flexural applications, and provides detailed guidelines for end anchor design, as well as lap and splice detailing. Updates also refine the treatment of FRP-concrete interface strain limits and incorporate comprehensive design examples for flexural and shear strengthening, as well as confinement.

The ACI 440.2R-17 guide provides a comprehensive framework for the design and application of externally bonded fiber-reinforced polymer (FRP) systems, covering flexural, shear, and confinement strengthening of concrete members. It introduces the environmental reduction factor, CE, which accounts for exposure conditions and long-term durability, supported by degradation models calibrated through experimental testing. The document also addresses key design considerations such as strain compatibility, potential bond failure modes, anchorage detailing, and serviceability requirements, including crack width and deflection limits.

The ACI 440 suite of documents provides a robust foundation for the design, specification, and implementation of externally bonded FRP systems for strengthening concrete structures. ACI 440.2R-17 serves as a comprehensive guide, outlining the principles of FRP behavior, durability considerations, and detailed methodologies for flexural, shear, and confinement strengthening. It introduces an environmental reduction factor CE, and long-term degradation models based on durability testing, while also addressing strain compatibility, bond failure modes, anchorage detailing, and serviceability limits such as crack width and deflection control.

Complementing this guidance, ACI CODE-440.13-24 (Strengthening Structural Concrete with Fiber-Reinforced Polymer (FRP) Systems—Code Requirements and Commentary) establishes mandatory code requirements for FRP strengthening, translating research and best practices into enforceable provisions for structural design professionals. Meanwhile, ACI SPEC-440.12-22 (Strengthening of Concrete Structures with Externally Bonded Fiber-Reinforced Polymer (FRP) Materials Using the Wet Layup Method—Specification) provides a standardized specification specifically for the wet layup method of externally bonded FRP application, covering material qualifications, surface preparation, installation procedures, and quality control measures. Together, these documents offer a cohesive and authoritative framework for ensuring the structural performance, durability, and constructability of FRP strengthening systems across a wide range of applications.

Additionally, the NCHRP Report 564 documents in-service performance of FRP bridge decks and highlights common field inspection methods and deterioration modes. Meanwhile, NCHRP Report 907 presents analytical and experimental results for prestressed concrete beams strengthened with externally bonded CFRP laminates, including long-term fatigue and creep behavior under sustained loads.

### 2.1.2.2 Europe

In Europe, external FRP strengthening is guided primarily by performance-based recommendations rather than prescriptive codes. The fib Bulletin 14 (2001) introduced fundamental concepts of externally bonded and NSM FRP systems, covering material selection, installation methods, anchorage techniques, and analytical modeling of strengthened members. This was later expanded by fib Bulletin 90 (2019), which provides advanced guidance on strength, serviceability, fatigue, fire resistance, and durability of externally strengthened elements.

While no harmonized European code exists specifically for FRP strengthening, design is often aligned with Eurocode 2 principles, using CEN/TS 1992-4-1 for adhesive anchors as a basis for bonded laminates. These documents stress the importance of environmental conditions, fire exposure, and creep under sustained loads. Partial safety factors are used in conjunction with reduction factors derived from exposure classifications.

#### *2.1.2.2.1 Country-Specific Codes in Europe*

Italy:
Italy has been a pioneer in FRP standardization through the National Research Council (CNR): CNR-DT 200/2025—Guidelines for the Design, Execution, and Inspection of Structural Strengthening Interventions Using Fiber-Reinforced Composites: The CNR-DT 200/2025 guidelines offer a comprehensive framework for the design, execution, and inspection of structural strengthening interventions using fiber-reinforced composites. Building upon previous editions, this document

integrates the latest advancements in materials science and structural engineering to address the rehabilitation needs of reinforced and prestressed concrete, as well as masonry structures. It encompasses detailed provisions on material characterization, design principles, execution methodologies, and quality control measures. Notably, the guidelines emphasize the importance of durability, providing considerations for environmental exposure and long-term performance.

Germany:
DAfStb Guideline "Strengthening of Concrete Structures with Bonded Reinforcement" (2012) focuses on bonded reinforcement using adhesives, including CFRP plates. While not FRP-specific, it applies to CFRP retrofits.

Russia:
GOST R 55243-2012—Composite Polymer Rebars for Concrete Structures establishes mechanical and durability requirements for GFRP/BFRP bars.

Supplemented by design manuals from NIIZhB and MGSU, using ISO/fib frameworks for structural evaluation.

#### 2.1.2.3 Asia

The Japan Prestressed Concrete Institute (JPCI) published its Recommendation for Design and Construction of Concrete Structures Using Fiber Reinforced Polymer (FRP) Composites in 1996. The latest version of the document was issued in 2021; it includes external strengthening methods for concrete structures using CFRP, AFRP, and hybrid systems, offering analytical models calibrated with Japanese bridge applications. The guidelines include strain limits, anchorage requirements, and compatibility checks for RC and PC members.

### 2.1.3 Material Specifications

The performance of FRP-reinforced or strengthened concrete elements relies heavily on the consistent quality and mechanical reliability of the composite materials used. To ensure safety, durability, and long-term serviceability, international material specifications define the minimum performance criteria and testing procedures for both FRP bars (for internal reinforcement) and FRP laminates (for external strengthening).

For FRP rebars used as internal reinforcement, standards typically specify tensile strength and modulus, in-plane shear strength, bond behavior with concrete, long-term performance under sustained stress (creep rupture), durability under environmental exposure, and thermomechanical limits such as glass transition temperature. Dimensional tolerances and product classification (e.g., GFRP, BFRP, CFRP) are also included to ensure compatibility with design specifications.

For FRP laminates or laminates used in external strengthening applications, material properties focus on in-plane tensile and shear capacity, bond performance to concrete substrates, environmental resistance, fatigue and creep behavior, and thermal stability. These parameters are critical to evaluating the contribution of

**TABLE 2.1**
**Comparison of material test standards for FRP composites**

| Property/Test Description | ASTM Standard | ISO Standard |
|---|---|---|
| Tensile strength and modulus of FRP bars | ASTM D7205 | ISO 10406-1:2015, Annex A |
| Tensile strength of FRP laminates/ laminates | ASTM D3039 | ISO 527-4, ISO 527-5 |
| Shear strength (in-plane) of FRP laminates | ASTM D5379 or ASTM D3846 | ISO 14129 |
| Bond strength of externally bonded FRP systems | ASTM D7705 | ISO 10406-2:2015, Annex B |
| Tensile modulus of elasticity of pultruded bars | ASTM D7205 | ISO 10406-1:2015 |
| Creep rupture and long-term stress durability | ASTM D2990, ASTM D7290 | ISO 10406-1:2015, Annex C |
| Environmental durability (exposure + residual properties) | ASTM D7705, ASTM D7290, ASTM D7957 | ISO 10406-1:2015 (Durability Classification) |
| Glass transition temperature (Tg) | ASTM E1356 | ISO 11357-2 |
| Specification for solid GFRP bars for internal reinforcement | ASTM D7957 / D7957M<br>ASTM D8505 / D8505M | ISO 10406-1:2015 |
| Specification for CFRP laminates and laminates for external use | ASTM D5117, ASTM D7565 | ISO 10406-2:2015 |

externally bonded systems to flexural, shear, or confinement upgrades, as well as their durability over time in various exposure conditions.

The two primary bodies publishing FRP material standards are ASTM International (commonly referenced in AASHTO and ACI documents) and the International Organization for Standardization (ISO). Table 2.1 summarizes key testing standards across both systems.

## 2.2 INSTALLATION PRACTICES OF FRP MATERIALS

FRP concrete elements can be divided into two major categories, distinguished as FRP reinforced or strengthened (internal or external installation) concrete elements, and can be subdivided by their application as well as accessibility, as it concerns the inspection requirement.

### 2.2.1 Internal Application (Reinforcing)

Concrete with embedded FRP bars or strands (EMBS), FRP can be used as bars or pre-tensioning/post-tensioning strands or bars. Beams, deck panels, columns/piers, pile/pier caps, piles, and footings can be reinforced by FRP bars or prestressing strands/bars. The installation practices vary depending on its use within the structural element.

#### 2.2.1.1 FRP Bars and Grids

AASHTO LRFD Bridge Design Guide Specifications for GFRP-Reinforced Concrete [111] provides construction specifications for cast-in-place structural elements. The guide specifies that GFRP rebars should be delivered, stored, and handled with care to prevent bending, coating with earth, oil, or other materials. Additionally, rebars must not be dropped or dragged and should be stored above the surface of the ground. Exposure to temperatures above 120°F during storage should be prevented. Proper construction practices and resin additives can mitigate the exposure to ultraviolet rays significantly. AASHTO highly recommends that, before placement in concrete, FRP bars are protected from direct exposure to sunlight and moisture.

The guide allows the use of steel tie wires, bar chairs, supports, or clips fully coated with epoxy or plastic to prevent displacements and maintain cover distances. A minimum concrete cover is required for GFRP rebars ranging between 1.5 in [38 mm] and 2.0 in [50 mm].

If bent bars are required, the bending shall be incorporated during manufacturing. Field bending is not permitted (for factory-formed bends, the minimum inside bend diameters shall conform to the requirements listed). All construction details should be shown on the project drawings. When GFRP rebars need to be cut in the field, shear or flame cutting is not permitted. In general, installation procedures are similar to steel rebars following the particular provisions in the guide.

### 2.2.2 External Application (Strengthening, Retrofitting, and Repair)

External applications are generally focused on strengthening, retrofitting, or repair of a steel-reinforced concrete element and intended to improve the existing capacity and performance. These objectives can be accomplished with externally bonded FRP laminates (EXBS) or near-surface-mounted FRP bars (NSMB).

#### 2.2.2.1 Externally Bonded FRP Laminates/Jackets

The externally bonded FRP laminates are normally used for strengthening by applications of one or more layers of FRP (laid or wrapped) to the surface of the concrete elements, and their application requires evaluation of the base material and surface preparation.

According to the Guide Specifications for Design of Bonded FRP Systems for Repair and Strengthening of Concrete Bridge Elements [112], "procedures for the installation of FRP systems are developed by the manufacturer and can vary between different systems". However, two different systems are stated in the guide as frequently used: dry fiber-fabric laminates saturated with an epoxy resin, referred to as the wet-layup technique (more versatile because the laminate is flexible and can conform to most shapes in the field) and the pre-cured fiber/resin laminates (manufactured and fabricated), which are bonded to the concrete surface with an adhesive resin.

To ensure a proper bonding between the concrete surface and the FRP system, the concrete surface should be dry and exhibit an open pore structure. AASHTO requires that the concrete surface be prepared to a minimum concrete surface profile (CSP) 3 as defined by the ICRI-surface-profile chips (referenced as ICRI 03732, NCHRP Report 609). Out-of-plane variations, including form lines, should not exceed 1/32 in. or the tolerances recommended by the manufacturer. All bond-inhibiting materials should be removed from the surface before the FRP application. All holes and voids should be filled with a compatible epoxy paste (thickened epoxy). When fibers are wrapped around corners, they should be rounded to prevent stress concentrations in the FRP system and voids.

#### 2.2.2.2 Near Surface Mounted (NSM)

This method involves carving grooves on the concrete surface and embedding FRP bars using grout or resin. NSM can be used for both strengthening and establishing connections. These systems consist of circular or rectangular bars or plates installed and bonded into grooves made on the concrete surface. A suitable adhesive is used to bond the FRP bar into the groove and is cured in place. ACI PCR440.2-23 [113] establishes the installation procedure following the system manufacturer. The two common FRP bar types used for NSM applications according to the guide are pultruded round and rectangular bars.

Grooves should be dimensioned to ensure adequate adhesive around the bars. Chapter 14.3 of ACI 440.2 defines 1.5 times the diameter of the FRP bar as the minimum dimension of the groove. However, when a rectangular bar with a large aspect ratio is used, the minimum groove size of 3.0 times the smallest side by 1.5 times the longest side is suggested. NSM systems can be used on the topside of structural members and for overhead applications. Adhesive type and installation method should be specified by the NSM system manufacturer.

# 3 Serviceability and Durability Criteria for FRP-Reinforced and FRP-Strengthened Concrete

When designing an FRP-reinforced concrete structural component, the minimum requirements of strength, stability, serviceability, durability, and integrity must be satisfied. Serviceability is defined as the ability of the structural system or structural member to provide appropriate behavior and functionality under the actions affecting the system, and it addresses issues such as deflections and cracking, among others [114]. Moreover, durability is defined as the ability of a structure or member to resist deterioration that weakens performance or limits the service life of the structure in the relevant environment considered in design [114]. In general, the strength and stiffness of the FRP composite may change under environmental conditions that could include water, alkaline or acid solutions, saline solutions, elevated temperature, and ultraviolet light exposure, therefore, potentially affecting the strength, serviceability, and durability of FRP-RSC elements.

In many instances, serviceability criteria or fatigue and creep rupture limit states may control the flexural design of concrete members reinforced with FRP bars, especially aramid FRP (AFRP) and glass FRP (GFRP) that exhibit low stiffness [115]. These limit states could be affected by environmental conditions that impact the mechanical and physical properties of the FRP bars. The deterioration of fibers, resin, and their interface governs the durability of FRP [116]. Tensile and bond properties of FRP bars are the primary parameters of interest for their use in reinforced concrete construction. In strengthening systems, the mechanical properties of externally applied FRP (i.e., tensile strength, ultimate tensile strain, and elastic modulus) may also be affected when exposed to environmental conditions such as alkalinity, saltwater, chemicals, ultraviolet light, high temperatures, high humidity, and freeze–thaw cycles [5, 69]. However, the serviceability limits of the FRP externally strengthened element are also limited by the original design of the steel RC element [117]. Figure 3.1 summarizes the physical, mechanical, and environmental factors that can influence the properties of FRP material. These factors will be discussed in the following along with a comparison between relevant serviceability criteria for the conventional steel-reinforced concrete and FRP-reinforced or FRP-strengthened concrete.

DOI: 10.1201/9781003646969-3

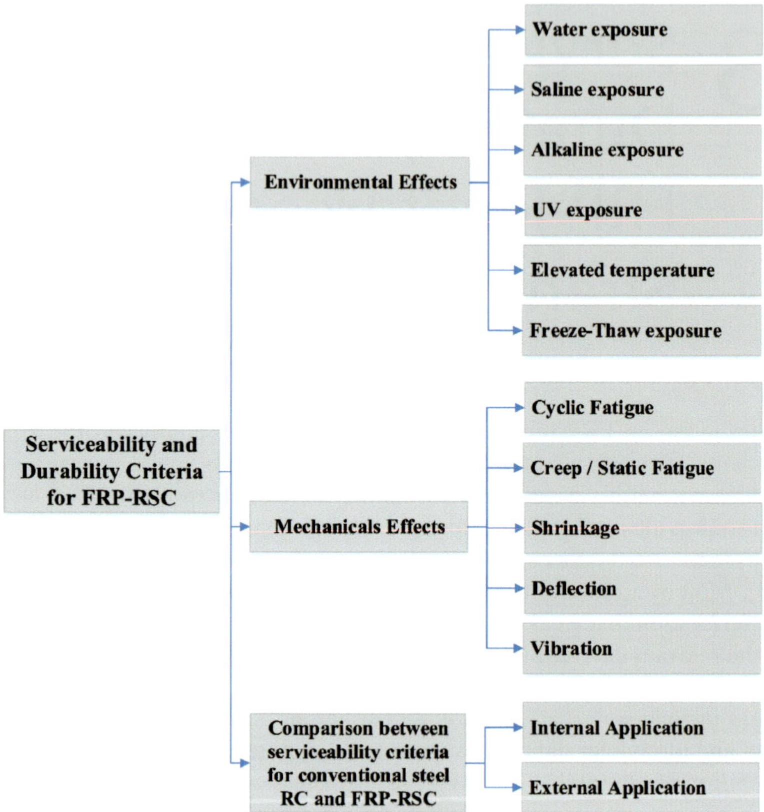


**FIGURE 3.1** Flowchart representing mechanical and environmental factors influencing the properties of FRP material

Exposures to conditions such as chemical contact, abrasive conditions, alkali-aggregate reactions, or other durability considerations pertinent specifically to the concrete performance that do not have a direct impact on the FRP composite behavior are only briefly described. These effects remain the same as those of steel-reinforced concrete. ACI 202.1R-16 [118] describes specific types of concrete deterioration mechanisms. Concrete, as a structural material, may require different degrees of durability depending on the exposure environment and desired properties. This chapter explains the consequences of different exposures and mechanical effects in the FRP composite, concrete, or FRP-RSC element.

## 3.1 ENVIRONMENTAL EFFECTS

Different environments could affect the properties of the FRP materials as well as the concrete embedding or supporting them. Changes in properties depend on the specific material and the exposure conditions. Nevertheless, due to the variety in fibers (e.g., glass, carbon, aramid, or basalt), resins (e.g., polyester, unsaturated polyester, epoxy, vinyl ester, modified vinyl ester, polyethylene), fiber volume fractions, solution concentrations, and temperature, among others, it is not appropriate to stipulate a single range of deterioration even for a specific composite (e.g., E-glass/vinyl ester) subjected to a specific harsh environment [119].

Current research on the environmental effects on FRP bars is mainly concerned with GFRP and BFRP bars [120]. Water, salt, acid, and alkaline solutions are commonly used to simulate conditions related to high humidity, seawater and deicing salt ambient, acid rain, and pore solution of concrete, respectively. Among these elements, the degradation of tensile strength of GFRP bars exposed to an alkaline solution has been reported as the worst [121]. In general, carbon fiber-reinforced polymer (CFRP) bars have better chemical resistance than GFRP and BFRP bars. When GFRP bars are immersed in water, the polymer matrix absorbs the water (or alkaline in the water), resulting in delamination and cracks at the fiber–matrix interface, which accelerates the diffusion of water and chemicals [122]. Degradation of glass fibers is associated with etching and leaching actions from moisture and alkalis and results in hydrolysis (breaking and weakening of bonds in polymer chains), plasticization (softening of the resin and hence low stiffness), and swelling, all resulting in degradation of the polymeric matrix [123]. Polymer composites exposed to a solution degrade at a rate that is related to the rate of absorption of the fluid, which in turn depends on the type of fluid, chemicals concentration, temperature, applied stress, damage status, chemical structure of the matrix, and fiber–matrix interface [43]. The interface between the fibers and the matrix is affected by these adverse conditions, hence introducing the potential for reducing the tensile properties of the GFRP composites. Moreover, aramid fiber-reinforced polymer (AFRP) bars, though often valued for their high tensile strength, are particularly susceptible to water absorption and ultraviolet (UV) degradation, both of which can significantly impair their long-term performance and durability.

### 3.1.1 Water Exposure

Exposure of FRP materials to water, high moisture, or high humidity can cause degradation in all three component levels of the composite: the fibers, the matrix, and the fiber–matrix interface [119]. Fibers that absorb water (aramid fibers and other organic fibers) undergo fiber swelling as a result of the water absorbed, consequently reducing their mechanical properties. However, the fibers that do not absorb water (carbon, glass, and basalt fibers) [124] may still experience strength reduction with the progression of microcracks that appear on their surfaces due to moisture ingress through the fiber–matrix interface, which accelerates the diffusion of water and chemicals [122]. The resin matrices, on the other hand, undergo deterioration due to plasticization and hydrolysis. Usually, water penetrates the resin matrix through diffusion and capillary action [125]. Plasticization softens the resin and hence lowers the stiffness of the FRP, while hydrolysis breaks and weakens the bonds in polymer chains, which is irreversible [126, 127]. Similarly, ingress of water along the fiber–matrix interface can cause degradation of the FRP material in two ways: swelling and dissolution. The water present at the interface between the fiber and the matrix causes the formation and propagation of cracks due to differential swelling. And also, the matrix material may dissolve into or leach away along the water absorbed at the interface, thereby degrading the fiber-to-matrix bond. Byars et al. [128] have stated that CFRP performs better than both AFRP and GFRP in moist conditions. Lebedev et al. [129] have mentioned that the resistance of BFRP to moisture effects is not inferior to that of GFRP, and occasionally it is even higher.

Tap water immersion is commonly used to simulate these conditions. A study carried out by Kim et al. [123] on GFRP/vinyl ester bars showed that delamination and some fiber degradation were caused due to water absorption by the resin. The tensile strength retention was about 88% and 80% after 132 days immersed in tap water at room temperature and 40°C, respectively. The elastic modulus was reduced by up to 25%, and the interfacial shear strength loss (to evaluate the degradation of interface between fiber and matrix) was about 20%.

Al-Salloum et al. [130] observed better residual tensile strength on the bars tested compared to most of the available literature at the time, noticing the improvement of the new generations of GFRP bars. Specimens in tap water at room temperature showed a slight reduction in the tensile strength of about 5% after 18 months of exposure. The presence of elevated temperature impacted the diffusivity which accelerates the tensile strength deterioration [123]. A strength loss of 25% was recorded when the temperature increased to 50°C. About 70% of the total loss occurred in the first six months of exposure. Tensile strength degradation of basalt fiber-reinforced polymer (BFRP/epoxy resin) bars varied from 3% to 38% when water temperature increased from 20 to 60°C [131]. Liu et al. [119] summarized the normalized residual tensile strength of pultruded FRP materials exposed to water/moisture based on experimental studies from 1996 to 2018. Figure 3.2 shows the results where only a trend of degradation can be observed because of the highly dispersed data.

Test results on GFRP (glass fiber types include E-glass and AR-glass) and BFRP bars immersed in tap water from previous studies were summarized by Duo et al. [120]. The studies include [44, 123, 130, 133–138]. In general, the strength retention ratios of both GFRP and BFRP bars gradually decrease with the increase in exposure time. From the studies collected, BFRP bars presented better deterioration resistance than GFRP bars. Table 3.1 summarizes the results of the test on tensile strength loss for GFRP bars exposed to water solution, while Table 3.2 shows the results of the test on tensile strength loss of BFRP bars exposed to water solution from various studies. It is worth noticing the differences in the results from different studies in relation to parameters such as fiber type, fiber volume fraction, polymeric matrix, bar diameter, and bar surface treatment that could impact performance.

#### 3.1.1.1 External Applications

In external applications, the polymer matrix (typically epoxy resin) is used as a primer, putty coat, and saturant. The surface of the element is primed with resin to act as an adhesive between the FRP system and the concrete surface. The presence

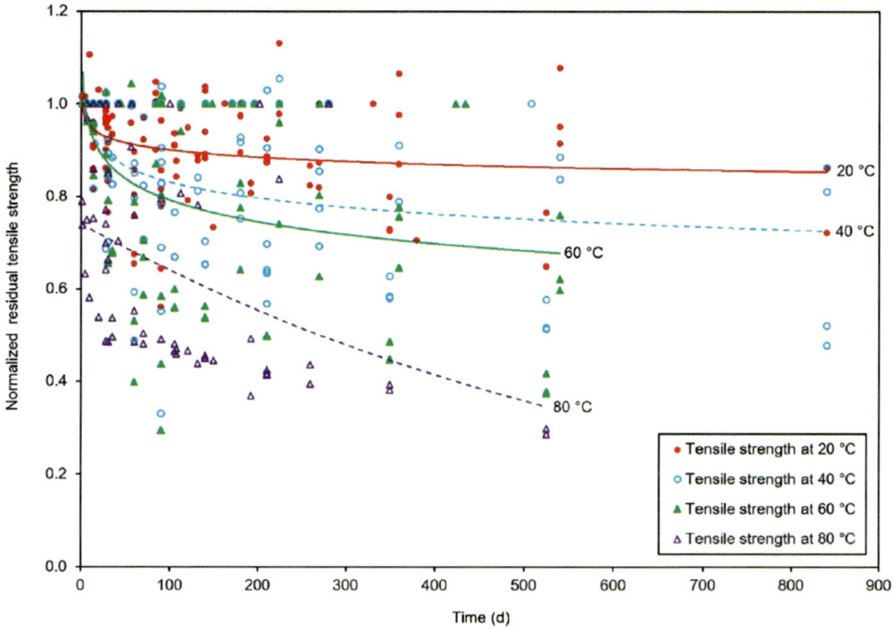


**FIGURE 3.2** Normalized residual tensile strength versus exposure time [132]

of humidity or water during installation or the service life of the element could be detrimental to epoxy resin due to its water absorption capacity that leads to a reduction in its stiffness, tensile strength, and its glass transition temperature [139, 140]. The epoxy resin is also commonly used to bond the FRP plies together when more than one layer is used. This interface between the adjacent layers is more susceptible to degradation by water [141, 142]. As the water exposure time increases, the failure mode of the strengthened element could change from fracture inside the substrate to adhesive failure. Sciolti et al. [142] observed that the glass transition temperature, stiffness, and strength of the epoxy resin are reduced when immersed in water owing to the plasticization effects.

### 3.1.1.2 Concrete

Concrete is resistant to most natural environmental conditions and chemicals. However, acids attack concrete by dissolving the cement paste and calcium-based aggregates. Using concrete with low permeability and treatment of surfaces may keep the concrete protected from these aggressive substances. Further, high amounts of sulfates in soil or water can attack and degrade concrete that is not properly designed [144]. Sulfate ions can react with calcium hydroxide and calcium aluminate hydrate to form gypsum and ettringite [145] and induce sufficient pressure to slowly cause the disintegration of the concrete. Concrete exposed to wetting and drying cycles is more affected by sulfates than in locations with continuous wet cycles.

**TABLE 3.1**
**Tensile strength loss of GFRP bars exposed to water solution in various studies**

| Reference | Matrix Material and Fiber Fraction | Diameter (mm) | Bar Surface | Solution | Temp (°C) | Exposure Period (h) | Tensile Strength Loss (%) |
|---|---|---|---|---|---|---|---|
| [133] | AR-glass (72%)/polyester | 10 | Ribbed | Water | 20 | 4320 | 7 |
| | | 19.5 | | | | | 6.4 |
| | AR-glass (72%)/ vinylester | 10 | Ribbed | | 20 | 4320 | 5.9 |
| | | 19.5 | | | | | 5.9 |
| [143] | E-glass/urethane-modified vinylester | 9.5 | – | Water | 50 | 4320 | 2.7 |
| [44] | E-glass (56%)/vinylester | 9.53 | Helically wrapped + sand coated | Water | 20 | 2880 | 5 |
| | | | | | 40, 60 | 1680 | 3, 29 |
| [123] | E-glass (73%)/vinylester | 12.7 | Helically wrapped + sand coated | Water | 25, 40, 80 | 720 | 8, 11, 15 |
| | | | | | | 1440 | 14, 15, 24 |
| | | | | | | 2160 | 20, 17, 21 |
| | | | | | | 3168 | 11, 16, 22 |
| | E-glass (69%)/modified vinylester | 12.7 | Sand coated | | 25, 40, 80 | 720 | 5, 8, 34 |
| | | | | | | 1440 | 24, 13, 45 |
| | | | | | | 2160 | 18, 22, 51 |
| | | | | | | 3168 | 12, 19, 56 |
| [130] | E-glass (83%)/vinylester | 12 | Ribbed | Tap water | Room, 50 | 4320 | 3, 17 |
| | | | | | | 8640 | 2.4, 22 |
| | | | | | | 12960 | 5.4, 25 |
| [121] | E-glass (64%)/vinylester | 8 | Grooved | Tap water | Room | 1080 | 2.7 |
| | | | | | | 2160 | 10.3 |
| | | | | | | 3240 | 14.7 |

**TABLE 3.2**
**Tensile strength loss of BFRP bars exposed to water solution in various studies**

| Reference | Matrix Material | Diameter (mm) | Bar Surface | Solution | Temp (°C) | Exposure Period (h) | Tensile Strength Loss (%) |
|---|---|---|---|---|---|---|---|
| Yu 2011 | Basalt (70%)/ unsaturated polyester | 8 | Helically wrapped | Water | 60 | 144 | 6.3 |
| | | 10 | | | | 720 | 6.5 |
| | | | | | | 1440 | 15.6 |
| | | | | | | 2160 | 15.9 |
| Wu et al. 2011 | Basalt (70%)/epoxy | 8 | Helically wrapped + sand coated | Water | 25 | 168 | 0.6 |
| | | | | | | 336 | 1.9 |
| | | | | | 40, 60, 80 | 168 | −3.4, 8, 23 |
| | | | | | | 336 | 3, 11, 19 |
| | | | | | | 720 | 7, 29, 41 |
| | | | | | | 2160 | 21, 31, 46 |
| [131] | Basalt (75%)/epoxy resin | 3 | Thin spiral ribs covered with a 0.5 mm sand coat | Water (pH = 7) | 20, 60 | 1000 | 3, 38 |
| | | | | | 60 | 200 | 18 |
| (Wu et al. 2015) | Basalt (65%)/epoxy | 6 | Helically wrapped | | 40 | 1008 | 2.8 |
| [121] | Basalt (64%)/ vinylester | 8 | Grooved | Tap water | Room | 1080 | 3.3 |
| | | | | | | 2160 | 7.1 |
| | | | | | | 3240 | 11.1 |
| | | | | | | 4320 | 11.5 |

### 3.1.2 Saline Exposure

FRP composites used in civil structures are usually subjected to saline exposure from construction in a marine environment or due to the use of deicing salts during winter. This exposure can contribute to FRP degradation. Although the exact reasons for the degradation of compressive, inter-laminar, and flexural strengths of FRP composites under saline exposure are not known [146], Rege and Lakkad [147] have suggested that the diffusion of NaCl molecules or the presence of salts could strain or rupture intermolecular bonds in the matrix as well as in the fiber–matrix interface. Additionally, the combined effect of moisture and salt crystals under freeze–thaw conditions contributes to the degradation of FRP composites as a result of the formation and expansion of salt crystals in the moisture-induced micro-cracks at the matrix–fiber interface [148].

FRP bars immersed in an aqueous solution with different types of salts do not necessarily have a significant difference in strength and stiffness when compared to those in a solution without salt [149, 150]. Composite specimens with higher fiber volume fraction were found to absorb more seawater compared with the specimens of lower fiber volume fraction [151]. Duo and co-workers [120] summarized the studies on glass fiber-reinforced polymers (GFRP, glass fibers types include E-glass and AR-glass-alkaline resistance) and basalt fiber-reinforced polymer (BFRP) bars immersed in a salt solution. In total, results from 40 and 20 tensile strength tests on GFRP and BFRP bars, respectively, were collected from the available literature [44, 133, 135–138]. Different concentrations were adopted to simulate the effect of seawater and deicing salt condition, including 3%, 3.5%, and 7% by weight as well as saturated solution. The modulus of elasticity did not show a significant degradation (a maximum of 9% degradation on BFRP bars at 60°C). The tensile strength degradation was about 12% and 18% for GFRP and BFRP bars at 60°C, respectively. Of the 20 BFRP bars tested, 8 BFRP bars were embedded in moist concrete and then immersed in the seawater. Severe degradation was reported in these tests which could be caused by the coupling action of the salt and alkali environment [123]. There was no significant difference between the results for AR and E-glass FRP bars.

Another study by Lu et al. [121] on E-glass FRP, BFRP, and CFRP immersed in simulated seawater for different exposure periods (a total of 4320 h) in water at room temperature with a pH of 7.1 according to ASTM D665-14e1 showed tensile strength retention around 81%, 75%, and 78% on E-glass FRP, BFRP, and CFRP, respectively. The study also indicated that the seawater solution caused greater moisture absorption than the alkaline and tap water solutions. Wang et al. [152] studied the long-term performance of BFRP and E-GFRP bars in seawater and sea sand concrete (SWSSC) environment subjected to accelerated corrosion tests using two types of SWSSC solutions at different pH and temperatures. Their study showed that GFRP bars had much better durability than BFRP bars, especially at high temperatures. This was attributed to the resin property of GFRP that is more stable in SWSSC environment than that of the BFRP bar. Morales et al. [153] carried out a study on unstressed GFRP/vinyl ester bars embedded in concrete mixed with seawater for up to 24 months of exposure in conditions of a subtropical region (25°C/71.2% RH). Based on the results, an exponential degradation model resulted in a long-term prediction (ten years) of the tensile strength capacity of 92% and 72%, for typical field exposure and aggressive exposure (seawater at 60°C), respectively. Other properties were measured (tensile strength, longitudinal elastic modulus, transverse shear strength, and apparent horizontal shear strength), but the tensile strength was the most affected. There have been limited studies on AFRP bars. A three-year exposure study evaluated pretensioned beam elements in a marine environment exposed to wet/dry cycles in saltwater to simulate tide change [154]. The results showed deterioration of bond between AFRP and concrete in ultimate load tests conducted on all the specimens exposed for two or more years. The ultimate capacity reductions were only 3.5% after three years.

FRP strengthening systems applied externally in marine environments or cold regions where salt deicing is used may go through a significant reduction in bond strength between the FRP and the bonded surface [69]. The durability of the FRP system can be strongly influenced by the resin properties, particularly in GFRP materials. The laminates are vulnerable to moisture due to the high absorption properties and the fact that most resins absorb between 1% and 7% moisture by weight. Other studies [155] showed that GFRP behavior degrades within a short time of exposure (1000 h) when it is in permanent contact with salty solutions. Silva and Biscaia [156] found that the first 1000 h of exposure to a saline environment did not significantly affect the FRP material, whereas at an exposure of 10,000 h (about 13 months), the interface at the FRP–concrete surfaces was affected significantly. Tests on the effects of saline water immersion on plates of GFRP showed that it can lower the mechanical properties by 17% after 3000 h of immersion (four months).

On the other hand, CFRP systems do not show considerable deterioration in saline environments [155]. CFRP is durable in typical field environments when compared to GFRP and AFRP materials. Chajes et al. [157] showed a 36% decrease in ultimate strength for GFRP retrofitted specimens that were subjected to 100 wet/dry cycles, while a 19% reduction was shown for CFRP-bonded specimens with a similar configuration. Toutanji and Gomez [158] observed a debonding failure mode that took place near the FRP–concrete interface. They reported a strength reduction of up to 33% on specimens made of different epoxies and subjected to 300 wet/dry cycles in saltwater.

Sulfates and chlorides in seawater require the use of low-permeability concrete to minimize steel corrosion and sulfate attack [144]. In steel-reinforced concrete elements, the rebars are protected from corrosion through the highly alkaline nature of the cement paste. The high pH environment in concrete (usually greater than 12.5) causes a passive protective oxide film on the steel. However, chloride ions from deicers or seawater can destroy this film and corrode the steel rebar.

### 3.1.3 Alkaline Exposure

Concrete itself is highly alkaline, which has implications for the durability of FRP composites. Carbon fibers are resistant to alkali and do not degrade as much as glass and basalt fibers [159]. In comparison to glass fibers, aramid fibers are less affected by alkali attack [128]. Silica, the major constituent of both basalt and glass fibers, is susceptible to chemical attacks and hence defines the chemical degradation mechanisms of both BFRPs and GFRPs [159]. The hydroxyl ions from an alkaline environment degrade the silica network (Si-O-Si + $OH^-$ → [Si-OH] + $SiO^-$) and weaken the silica lattice bond in basalt and glass fibers [129, 159]. This leads to a gradual loss of weight and strength of the BFRPs and GFRPs under strong alkali attack. Microcracks generated on the surface of the polymer matrix when the GFRP rebar is stressed allow the alkaline solution to enter, ultimately leading to the fracture of the rebar [122]. Furthermore, the contact layer between the fibers and polymer matrix is damaged because of the chemical reaction, increasing the hydration product and reducing the mechanical performance of GFRP rebar.

Alkaline solutions with pH values varying between 11.5 and 13.0 degrade the tensile strength and stiffness of GFRP bars [160]. Li et al. [161] studied the durability and fatigue performance of BFRP (epoxy matrix) bars in an alkaline environment. The bare basalt fiber immersed in this environment, without resin protection, exhibited a serious degradation of tensile properties due to severe fiber corrosion, as revealed by scanning electron microscopy.

Benmokrane et al. [16] immersed basalt and glass FRP bars in an alkaline solution for up to 5,000 h at 60°C to study the durability of the bars made with vinyl ester and epoxy resin by assessing the transverse shear strength (major structural force on dowels in jointed pavements or on stirrups in concrete beams), the flexural strength, and the interlaminar shear strength (governed by the fiber–matrix interface). The best properties were found in the GFRP bars with vinyl ester, which had the lowest degradation rate. The BFRP bars with vinyl ester showed the worst degradation, with a transverse shear strength reduction of 33%. The strength degradation observed for both types of basalt FRP bars, according to the authors, was attributed to the sizing, which evidenced a poor bond between the resin and basalt fibers. Another study on BFRP bars with vinyl ester also affirmed that the damage resulting from the presence of alkali is controlled by the matrix at the fiber–resin interface when studying the interlaminar shear strength degradation [137]. Commonly used GFRP bars are mainly formed by E-glass (electrical glass) fibers impregnated with the resin matrix. The tensile strength losses of E-glass and ECR glass bars (chemical resistance E-glass, which have a higher content of silica) were 15–30% and 8%–18%, respectively, for 15-mm diameter bars and the same sizing agent (thin coating applied to the surface of the fiber) after the same exposure time in alkali condition [162, 163].

Duo et al. [120] summarized the studies on GFRP (E-glass, AR-glass-alkaline resistance, and ECR glass) and BFRP bars immersed in alkali solution. The fiber volume fraction ranged from 56% to 72%. The FRP bars were immersed in alkali solution with different concentrations (pH varying from 12 to 13.6) for different time periods, and heating was used to accelerate degradation progress [17, 44, 123, 133, 137, 138, 164–166]. The average strength retention ratio in alkali solution at 20°C was 89% and 91% for GFRP and BFRP bars, respectively. The tensile strength retention at 80°C was on average 60%. The modulus of elasticity did not show significant deterioration. Compared to elevated temperatures, at room temperature (20°C), the rate of degradation is slower with increasing exposure time. The BFRP bars with a smaller diameter had a similar degradation rate but maintained higher residual strength compared with the BFRP bars with a larger diameter in the same condition of pH and exposure time [137]. Unlike the steel bars, whose properties can be assumed to be the same for different diameters, the longitudinal tensile strength of FRP rebars is diameter- or size-dependent due to a phenomenon known as the "shear-lag" effect [10]. The alkaline solution with higher temperature significantly increases the degradation ratio of the FRP bars.

From the tests data collected by Duo et al. [120], all the ECR glass bars had a vinyl ester matrix and showed a better alkaline resistance than E-glass. When immersed in 20°C, 40°C, and 60°C alkaline solutions, the average strength retention ratio was found to be on the order of approximately 90–100% after about 6,000 h at 20°C, about 80–90% after about 4,000–6,000 h at 40°C, and about 70–80% after up to 17,280 h at 60°C. For E-glass on average, when immersed in 20°C, 40°C, 60°C, and 80°C alkaline solutions, the average strength retention ratio was found to be on the order of approximately 85–95% after about 5,000–6,000 h at 20°C, about 80–85% after about 4,000 h at 40°C, about 60–70% after about 5,000 h at 60°C, and about 50–60% after similar exposure durations at 80°C.

A study carried out by Benmokrane et al. [162] evaluated the durability of 24 types of FRP bars in a simulated alkaline concrete environment, specifically concerning FRP type (carbon, basalt, and glass), fiber sizing, resin chemistry, and manufacturer. The tests showed that the carbon fiber-reinforced polymer (CFRP) bars have the highest tensile-strength retention and resistance to the alkaline environment (rated very good with over 80% strength retention), followed by the basalt and glass bars. The average transverse-shear strength of the FRP bars studied (including E-glass, ECR-glass, BFRP, and CFRP) exceeded the specified limit of 124 MPa in ACI 440.6M-08 and 160 MPa in CAN/CSA S807-19, and there was no statistically significant difference between the average transverse-shear strength of the E-glass and ECR-glass bars, or between the BFRP and CFRP bars. Robert et al. [165] found that the mortar-embedded GFRP bars exposed to tap water were less affected by accelerated aging than the bars exposed to the simulated pore-water solution, showing the underestimation of GFRP durability and conservative design of concrete structures reinforced by composite materials.

An experimental study carried out on BFRP bars embedded in seawater-sea sand concrete (SSC) immersed in different solution environments showed that alkalinity is the main factor controlling the performance degradation of BFRP bars on the horizontal and transverse shear strengths [167]. The degradation of BFRP was strongly correlated with the concrete cover size; the greater the cover thickness, the greater the degradation. This could be attributed to a larger shear crack size in the case of larger concrete cover. Other tests after 4,320 h in water at room temperature with a pH of 12.9 (according to ACI 440.3R-04) showed tensile strength retention of around 77%, 82%, and 69% for BFRP, CFRP, and E-GFRP bars, respectively [121]. In general, no significant degradation of elastic modulus for the three types of FRP bars was found after exposure to alkaline solutions. The appearance changes of E-glass

FRP bars were more obvious, with some damage characteristics such as resin spalling and surface deterioration with blisters or particles.

Few studies have focused on aramid fiber-reinforced polymer (AFRP) bars compared to other fiber types. The AFRP tensile strength and stiffness in alkaline solutions at elevated temperatures for either stressed or stress-free conditions have been reported to decrease between 10–50% and 0–20% of the initial values, respectively [168]. Tests on the aramid bar also showed sustained short-term strength ratios of 75%, 70%, 60%, and 50% for exposure to 20°C air and 20°C, 40°C and 60°C alkaline environments, respectively (at 10,000 hours) [169]. In a study that evaluated pretensioned elements in a marine environment exposed to wet/dry cycles in saltwater to simulate tide change after three years of exposure, Sen et al. [154] concluded that concrete's alkaline environment has negligible effect on the AFRP material. The ultimate capacity reduction was only 3.5% compared to unexposed control specimens. Liu et al. [119] summarized the normalized residual tensile strength of pultruded FRP materials exposed to strong alkaline solutions, as shown in Figure 3.3. They stated that a higher alkalinity could induce more severe degradation in FRP materials than other environments.

As it relates to external FRP applications, Mohammadi et al. [170] immersed 24 CFRP laminate specimens in alkaline conditions for 3,000 h. In general, alkaline conditions led to a significant reduction in the bond strength. Results indicated that

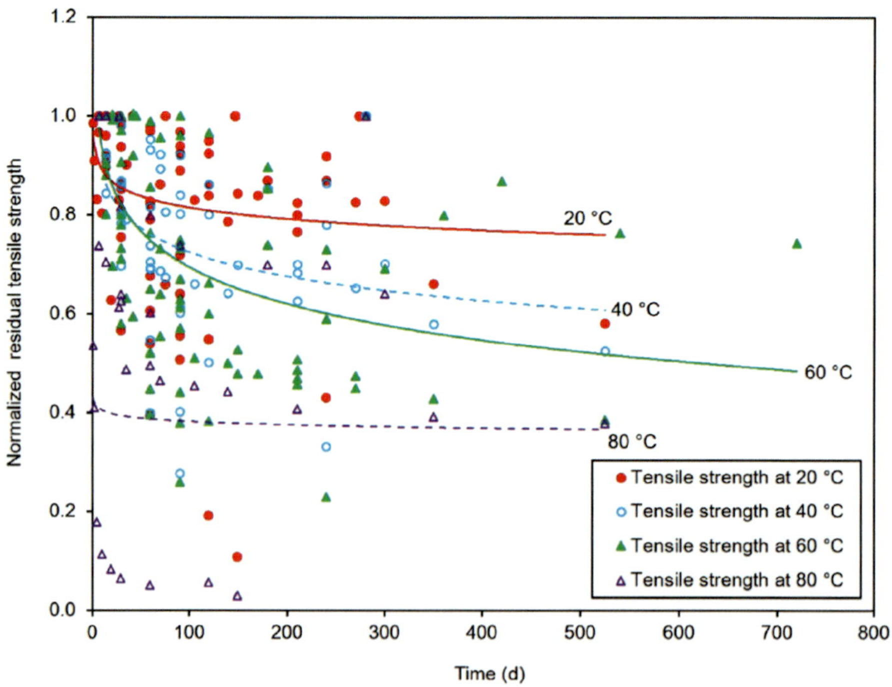


**FIGURE 3.3** Normalized residual tensile strength versus exposure time [132]

specimens exposed to an alkaline solution at 23°C experienced a reduction in their load-carrying capacity by 4.7%–23.3%. When the temperature increased from 23 to 60°C, the change in bond strength showed a slower rate of reduction. Test data on AFRP showed that strength can be reduced by 55% to 60% in alkaline solution [69].

In concrete, the alkali-aggregate reaction (AAR) between the alkalis in pore solution and some active chemicals from aggregates can produce significant expansion that could be detrimental. Two types of AAR have been recognized: Alkali-Silica Reaction (ASR) with active minerals from amorphous silica and Alkali-Carbonate Reaction (ACR) with active minerals from dolomitic limestone [118, 171]. Generally, the degree of AAR is influenced by the presence of water (ASR only occurs in high relative humidity), alkali content, concrete porosity, and temperature. A gel is formed when the silica present in aggregates reacts with the alkali hydroxide in concrete. It can swell when it absorbs water from the surrounding cement paste or from the environment and hence, the expansive pressure may produce damage to the concrete. Random map cracking, closed joints, and spalled concrete are typical indicators of ASR [144].

Due to the diffusion of carbon dioxide from the atmosphere and ingress of chloride ions, the natural alkalinity within the concrete is lost. In RC elements reinforced with steel, this can result in a reduction in pH with breaking the passive layer of iron oxide around the steel rebar, and consequently, the steel can start corroding if sufficient oxygen and moisture are present. This is detrimental to the structural capacity and serviceability of concrete elements. Carbonation has been found to have a positive effect on concrete strength, increasing the strength of carbonated concrete. Demis and Papadakis [172] found that until the carbonation reaches the FRP bar, no significant deterioration of the bond is evidenced. Overall, the expected bond deterioration due to carbonation is compensated by the increase in concrete strength for FRP RC elements.

### 3.1.4 Ultraviolet (UV) Light Exposure

The UV exposure may result in surface oxidation due to different chemical mechanisms related to the resin type [43, 173]. UV radiation is capable of degrading the molecular bonds in the polymer matrix and hence has damaging effects on the FRP [174]. The reason behind this is that the wavelength of the UV radiation (typically between 290 nm and 400 nm) is in the same range as the wavelengths of the bond dissociation energies of the polymers [175, 176]. Although the UV exposure can degrade only the top few microns of the surface, the damaged area could be the point of stress concentration and might further invite other environmental attacks. The FRP rebars embedded in concrete are themselves protected from the UV exposure, but they are vulnerable during storage or when used as external reinforcements [128, 177]. Ultraviolet stabilizers could be added to protect the resin from the effects of sunlight [178].

Some results from combined ultraviolet and moisture exposure tests have shown tensile strength reductions of 0% to 20% of initial values in CFRP, 0% to 30% in AFRP, and 0% to 40% in GFRP bars with and without stresses applied to them

[179, 180]. About 10-mm diameter AR-glass FRP bars (polyester and vinyl ester matrix) exhibited a tensile strength loss of about 5.5% when subjected to ultraviolet radiation at the rate of $31.7 \times 10^{-6}$ J/s/cm$^2$ simulating exposure to sunlight in a desert environment for a period of six months [133]. Tests on CFRP, AFRP, and GFRP rods showed a tensile strength loss after three years of exposure of about 23%, 31%, and 10%, respectively [64]. Additional studies according to [46] have shown a reduction in tensile strength for AFRP rods around 13% after 2,500 h exposure, 8% reduction for GFRP rods after 500 h (no reduction thereafter), and no reduction for CFRP rods. The use of AFRP bars is limited due to its sensitivity to UV radiation. Moreover, one of the advantages of BFRP bars is its resistance to UV light [168]. A study of 4-mm BFRP bars (vinyl ester matrix) showed no damage on the bar surface and 89% strength retention after UV exposure for 480 h at 50°C [181].

On the other hand, the effect of ultraviolet (UV) radiation is of vital importance for applying FRP composites as an external reinforcement [182, 183]. Previous results in AFRP laminate composites tests showed that both fibers and resin are affected by UV light. The strength retention of a 0.13 mm laminate after 1000h exposure was about 60% [184]. Glass and particularly carbon fibers are less sensitive to the effects of UV radiation, though the majority of resins will be affected by UV [168]. To prevent the effect of UV radiation, structural measures or material modifications (extra matrix additives, pigmented gel coatings, painting) could be used. According to the CSA S806 [185], UV exposure can cause embrittlement and micro-cracking in an unprotected laminate surface. Color shift or yellowing and gloss changes are some of the effects of UV exposure on FRP laminates. Gel coating protection is the most common protection method against UV because it provides a deep 10 to 20 mm thick protective layer. An experimental study on FRP laminates used as externally strengthening material [186] showed that the ultimate tensile strength and modulus of elasticity of FRP laminates decrease with UV exposure time, with the degradation of resin adhesive material being the main degradation parameter. They also found that the increase in the number of FRP layers does not help to reduce the UV exposure degradation and that CFRP had the most stable mechanical properties under this exposure.

### 3.1.5 Elevated Temperature Exposure

The deterioration due to thermal exposure such as temperature variations and high temperatures occurs as there are differences in the coefficient of thermal expansion (CTE) of fibers, resins, FRPs, and concrete [128]. Due to the anisotropic behavior of FRP bars, the coefficient of thermal expansion is different in longitudinal and transverse directions, the first one controlled by the fibers and the second one by the resin [187, 188]. The CTE along the longitudinal direction of FRP is similar to that of concrete, but it is 5–8 times higher along the transverse direction [177] which should be considered particularly in relation to bond behavior [128]. Due to differences in CTE, high temperature variations can introduce high outward pressure on the FRP reinforcements, causing longitudinal splitting of concrete if the confining action of the concrete cover is not sufficient. Further, elevated temperature results in micro cracks at the interface between FRPs and the concrete, which significantly decreases the FRP-substrate bond as well as the inter-laminar bond in the FRP [43, 175, 189]. The bond strength between the FRP bar and concrete at high temperatures is strongly related to the glass transition temperature ($T_g$) of the resin [190]. Similarly, under high temperatures above the glass transition temperature, the resin matrix softens and transitions into a rubber-like material, which reduces the matrix stiffness and degrades the fiber–matrix interface [119]. At temperatures above 300–400°C, thermal decomposition of the FRP organic matrix usually occurs with potential emission of smoke, soot, toxic/combustible volatiles, and heat. Organic fibers (e.g., Aramid) may also degrade and produce smoke, fumes, and heat [191]. While concrete generally has a high resistance to fire, its mechanical properties could also degrade when exposed to elevated temperatures. High temperature further accelerates the environmental effects (water, alkaline, UV exposure) discussed above.

Tests on GFRP and CFRP bars made with long (but not continuous) fibers and polyester resin showed that the stress-strain relationships of FRP bars remained almost linear at elevated temperatures until failure [192]. At temperatures of above 350°C, the resin begins to decompose. Before reaching 400°C, little impact on the elastic modulus was evidenced, retaining about 90%.

Other studies have shown that CFRP, GFRP, and AFRP materials experience a reduction in their tensile strength and bond properties at temperatures between 100°C and 350°C [193, 194]. Tensile tests carried out on ECR-glass FRP (vinyl ester hybrid resin) bars exposed to elevated temperatures indicate a limit for temperature ranging from 400 to 450°C, corresponding to a tensile strength loss of around 30% to 55% [195]. The bond strength between FRP bar and concrete showed a substantial decrease near 180°C (close to the glass transition temperature). Hajiloo and Green [190] carried out pull-out tests on 16-mm GFRP bars (sand-coated, sand-coated braided, and ribbed with an average of 84% fiber volume and Tg ≈ 120°C) under steady-state and transient temperature protocols (temperature ranging from 25 to 360°C). The tests showed that when the concrete-to-bar interface temperature reached 75°C the bond strength losses were around 27%.

Related to fire exposure, a study carried out by Kodur and Bisby [196] showed that CFRP and GFRP-reinforced slabs (no information about fibers and resin was provided) have lower fire resistance than steel reinforced slabs in terms of the critical temperature for the reinforcement (described as those at which the FRP bar lose 50% of their ultimate tensile strength), and are influenced by the aggregate type and concrete cover. After four hours exposure to fire, spalling was not observed in any of the slabs. They concluded that the heat transfer behavior of FRP–RC slabs appears similar to slabs reinforced with steel bars, and that the FRP-RC slabs have much lower fire resistance. However, this was based on assumed critical temperatures of 250° and 325° for CFRP and GFRP, respectively. The study did not account for thermally induced bond degradation [192].

Two full-scale GFRP (vinyl ester and polyurethane thermoplastic resin) reinforced concrete beams (350 × 400 × 4400 mm, concrete cover of 75 mm) exposed to fire and service load showed a minimum period of fire resistance of 90 minutes, for GFRP vinyl ester beam reaching up to 120 min [197]. A lap splice test on a 5 m by 2 m GFRP-reinforced concrete slab [195] exposed to fire according to DIN EN 1363 for 90 minutes (the element was subjected to service load which caused a stress in the rebar of 300 N/mm$^2$) showed a bond failure mode when the temperature at the centerline of the bars reached approximately 230°C.

Nigro et al. [198] tested four GFRP-reinforced concrete slabs under typical design loads exposed to fire action. Two more unloaded GFRP slabs were tested to evaluate the residual resistance. The results indicated that temperatures of up to 460°C may be reached with a failure associated with rupture of fibers in the middle of the element if anchorage length (i.e., 500 mm) outside the exposed zone is provided to avoid pull-out of the bar. This anchorage length ensures the resistance beyond the glass transition temperature (Tg) in the exposure zone when the resin softening reduces the adhesion between FRP and concrete. A bent end may reduce the anchorage length. However, the bar could still become unbonded and larger cracks could appear. If sufficient anchorage length is not provided, the fire endurance will depend on the slippage of the bar at temperatures of around Tg [198–200].

Six over-reinforced CFRP- and GFRP-RC beams tested under a four-point static load (the load kept constant during the test) at elevated temperatures achieved the critical bar temperature of 500°C set as failure criteria (literature suggested a critical temperature for FRP ranging between 250–500°C) [201]. All rebars showed similar temperatures independent of bar material, and the CFRP beam achieved the higher load and failed by concrete crushing (slip was registered after 45 minutes of exposure). The CFRP-RC beam showed less stiffness degradation compared to GFRP- and steel-RC. The rate of stiffness degradation on the GFRP-RC beams was almost independent of both load level and rebar modulus.

Table 3.3 shows the relevant results on studies related to FRP-RC elements exposed to elevated temperatures as discussed previously. The impact on the mechanical properties in FRP bars embedded in concrete exposed to fire or elevated temperature is quite evident. Extended bar anchorage and deeper concrete cover could improve the fire performance of the concrete element. However, the overall performance depends on several parameters related to the bar constituents.

Moderate ranges of temperature changes, around ±50°F (±28°C) have been shown not to affect the bond between surfaces of FRP strengthening elements [69]. First studies of the fire performance of RC beams strengthened in flexure with externally CFRP strips in Switzerland [205] observed a fire endurance of 81 minutes for unprotected specimens and 146 minutes for specimens with thermal insulation. Blontrock et al. [206] tested RC beams strengthened with externally bonded CFRP strips protected with calcium silicate boards and reported a loss of bond between surfaces within the first 40 minutes of exposure at temperatures close to the glass transition temperature of 55–60°C regardless of the insulation scheme used. A U-shaped fire protection system applied to both the base and sides of the beams was found to be most effective at prolonging the time before loss of interaction between surfaces.

A study in the performance in fire of CFRP-strengthened reinforced concrete slabs, beams, and columns [207] demonstrated that sufficient fire endurance can be provided with adequate insulation. The slab tests suggested that even with 38 mm of supplemental insulation, it will be very difficult to prevent FRP temperature

**TABLE 3.3**
**Research plan and results summary of the reported structural responses of FRP-reinforced concrete members exposed to elevated temperatures**

| Reference | Sample | FRP Type | $T_g$ (°C) | Exposure Condition | Period (min) | Results |
|---|---|---|---|---|---|---|
| [197] | RC Beam ($f'_c \approx 42$ Mpa), 75 mm concrete cover | GFRP (vinylester and polyurethane) | N/A | Fire (460°C and 380°C max in rebar) | 100–140 | Fire tests indicate fire design minimum requirements (90 min) based on BS 476 |
| [192] | 200 mm bar (9.5 mm and 12.7 mm diameter) | GFRP (polyester) | N/A | Temperature (100–600°C) | - | 0%–99% tensile strength loss, -17%–72% modulus of elasticity loss |
| [196] | RC Slab (f'c ≈ 40 Mpa), 25.4 mm concrete cover | GFRP/CFRP bars | N/A | Fire (up to 800°C) | 240 | 50% tensile strength loss, a test fire endurance between 35–45 min |
| [195] | RC Slab 60 mm concrete cover | GFRP bars | N/A | Fire based on DIN EN 1363 | 90 | Bond failure mode when temp in the bars reached approximately 230°C. |
| [198] | RC Slab (f'c ≈ 39 Mpa), 32–51 mm concrete cover | GFRP bars (70%) | 100 | Fire based on ISO 834 | 240 | Cover values of 51 mm and anchorage length values of about 500 mm showed better structural behavior |
| [201] | RC Beam ($f'_c \approx 35$ Mpa), 20 mm concrete cover | GFRP/CFRP bars | N/A | Temperature (up to 900°C, 500°C max in rebar) | 80 | CFRP RC beams indicated better stiffness characteristics |
| [202] | RC Beam (hybrid concrete) | GFRP bars | N/A | Temperature (300–700°C) | 200 | GFRP-RC beam capacity was reduced by around 53% at 700°C. |
| [203] | RC Beam ($f'_c \approx 40$ Mpa), 30 mm clear cover | BFRP; Hybrid FRP; nano-Hybrid | N/A | Fire, up to 800°C | 80 | Reduction in the strength capacity by 43%, 40%, and 43% |
| [204] | RC Beam ($f'_c \approx 41$ Mpa), 20 mm concrete cover | GFRP/CFRP bars | 120 | Fire, up to 1000°C | 120 | The CFRP beam reached 66% of initial load bearing capacity. The 14-mm GFRP beam reached higher capacities than the 10 mm |

from exceeding the glass transition temperature of the resin for more than one hour. Another study on T-beams strengthened with externally bonded CFRP fabrics [208] demonstrated that one layer of VG insulation (vermiculite gypsum—a lightweight, fire-resistant cementitious plaster that can be spray-applied) of 25 and 38 mm can achieve fire endurance of more than 4 hours (based on ASTM E119), the Tg of the epoxy was exceeded between 16 and 36 minutes and between 55 and 57 minutes for the 25 and 38 mm of insulation, respectively. The latter kept the average FRP temperature below the matrix Tg for 54 minutes. Similarly, the concrete cover, owing to its low thermal conductivity, is essential for FRP bar protection when exposed to elevated temperatures [198]. Aiello et al. [209] stated that a cover value greater than two times the bar diameter could avoid the occurrence of the splitting cracks when subjected to a temperature increase of 50°C.

### 3.1.6 Freeze–Thaw Exposure

Freeze–thaw cycles along with the coupled effect of high moisture may lead to deterioration of FRP composites [119, 210]. The FRP materials become brittle under freezing temperatures and also the moisture absorbed expands when it freezes, leading to microcracks, particularly at the interface between the FRP and the substrate material [175]. The freezing and thawing cycle causes the degradation of the fibers, resin, and interfacial bond in the FRP composites as well as of the bond between concrete and FRP [177]. Exposure to freeze–thaw cycles could affect the FRP composites themselves due to mismatched coefficients of thermal expansion of constituent elements, which may cause micro-cracking and void generation [211]. Further, the interface between concrete and FRP bar could be affected as the difference in coefficients of thermal expansion generates radial pressure at the interface, and ice crystal growth during repeated freeze–thaw cycles generates micro-cracking due to the high pressure in the concrete, reducing the bond strength between them [212, 213]. These micro/macro-cracks could accelerate degradation by allowing the penetration of other chemical solutions.

Experimental studies on FRP bars have shown different performances when exposed to freeze–thaw cycles. Uomoto et al. [64] reported that strength losses for CFRP, GFRP, and AFRP rods subjected to 300 freeze–thaw cycles varying from –20 to 15°C were negligible. Pull-out test results of CFRP, GFRP, and AFRP bars (freeze–thaw cycles ranging from 50 to 600) [214] showed that the bond strength between FRP bar and concrete surface was not considerably influenced by the environmental condition. However, AFRP bar specimens lost about 40% bond strength after 600 freeze–thaw cycles.

In another investigation, FRP specimens (Carbon/epoxy-vinyl ester, carbon/epoxy, E-glass/thermoplastic, and E-glass/polyester) were subjected to environmental cycles to investigate how the combination of temperature cycles, high moisture, and UV radiation can affect the long-term behavior of the FRP bars [43]. The high temperature cycle was chosen to vary between 16 and 49°C to simulate the summer season effects. Freeze–thaw cycles varying between –18 and 4°C were considered as a simulation of winter effects in accordance with the ASTM C666-92. However, these environmental combined cycles did not produce significant reductions in the physical-mechanical properties (E-glass/polyester tensile strength reduced by 6.6%).

Won and Park [215] investigated the effect of 300 freeze–thaw cycles on the bond strength of CFRP and E-Glass FRP bars (vinyl ester, modulus of elasticity of 123 and 43 GPa) with helical wraps using pullout specimens. The results indicated a 20% decrease in the bond resistance of the CFRP and GFRP bars. E-glass (vinyl ester and modified vinyl ester) and Carbon (epoxy resin) FRP bars exposed to 250 freeze–thaw cycles from –29°C to 20°C [211] showed no more than 10% deterioration in tensile strength. Kim et al. [123] stated that tensile strength reduction of E-glass or vinyl ester FRP rods immersed in alkali solution (pH = 13) at different temperatures in both freeze–thaw + alkali and alkali exposure specimens was reduced mainly due to the alkali attack.

Results of testing on sand-coated GFRP-reinforced concrete (6% entrained air content) beams exposed up to 360 freeze–thaw cycles (–20 to 20°C), either in an unstressed state or loaded in bending, showed no significant effect on the behavior in terms of deflections, strains, and ultimate capacity for the two actions when compared to control specimens [216]. Some specimens showed even better resistance after the test, which may be explained by the increase in concrete strength during conditioning (50% humidity). Alves et al. [217] studied the individual and combined effects of freeze–thaw cycles along with sustained axial load and fatigue loading on the bond characteristics of the sand-coated GFRP/vinyl ester bars embedded in concrete. They showed that the effect of fatigue loading was more dominant than that of the freeze–thaw effect. Further, freeze–thaw cycles along with the sustained load effect increased the bond strength between GFRP bars and concrete by approximately 40%. They explain this by the fact that the GFRP bar absorbs the moisture and enlarges in cross-sectional area, enhancing the friction mechanism. Other study [212] on GFRP-reinforced concrete slabs subjected to sustained load (20% to 30% of flexural capacity) and freeze–thaw cycles (30 cycles varying from -30 to 30°C) showed that thermomechanical load applied before bending tests contributed to an increase in the shear ultimate strength of slabs for a ratio of concrete cover to bar diameter of between 1.6 and 2.8. However, the stiffness at service load was lower.

Freeze–thaw cycles can cause progressive degradation of the FRP composite, weakening the interfacial bond at the FRP to concrete interface [69]. The performance of RC elements strengthened with externally bonded FRP laminates exposed to freeze–thaw cycles is generally associated with either the bond between the FRP laminate and concrete or the mechanical properties of the laminates [218]. Chajes et al. [157] reported tensile strength loss for CFRP-strengthened beams of 9% when exposed to freeze–thaw cycles compared to either GFRP or AFRP, which showed losses of about 50%. Karbhari and Zhao [219] also reported a better performance of the CFRP over the GFRP strengthening system, stating that freeze–thaw cycles can cause matrix hardening, microcracking, and overall bond degradation. Other studies have reported no changes in the strength performance when exposed to freeze–thaw cycles. Mukhopadhyaya et al. [220] noted that the ultimate bond strength between FRP and concrete was not affected. The failure mode changed from concrete shear failure to adhesive failure after being exposed to freeze and thaw cycles. However, some studies have shown increases in performance, like Green et al. [221] who reported that single lap shear specimens exhibited ultimate strength increases of up to 54% after the freeze–thaw exposure, and failure mode progressively changed from concrete shear failure to adhesive failure.

Homam and Sheikh [222] immersed CFRP and GFRP coupons in water and then subjected them to up to 300 freeze–thaw cycles between –18°C and 4°C. Tensile strength was not significantly affected by the exposure, but the average bond strength of exposed GFRP specimens decreased with the increase in the number of cycles. RC beams strengthened with externally bonded CFRP and GFRP plates in flexure subjected to up to 200 freeze–thaw cycles [223] showed no significant adverse effect. The influence of freeze–thaw cycles on the CFRP–concrete interface fracture properties using direct shear tests was studied by Subramaniam et al. [224] who reported a 17% decrease in ultimate load. Soliman, El-Salakawy, and Benmokrane [225] reported a failure mode governed by adhesive material (epoxy) when NSM CFRP concrete specimens were subjected to freeze–thaw cycles. Hair cracks developed in the adhesive material caused a strength loss between 8% and 14%. A study on GFRP and CFRP coupons showed no significant influence on the mechanical performance when exposed to freeze–thaw cycles [218].

Damage in concrete due to freeze–thaw cycles depends on the saturation of the concrete. Freeze–thaw cycles have minimum adverse effects on dry concrete, even under a relative humidity of 75–80% [213]. The freeze–thaw damage to concrete may be affected by a complex interaction of numerous factors, including concrete permeability, temperature gradient, and air void systems. Deterioration could be accelerated particularly in the presence of deicing chemicals. Deterioration is caused by the freezing of water and subsequent expansion in the paste, the aggregate particles, or both [144]. Microscopic air bubbles within the concrete, through the addition of an air-entraining admixture and thorough mixing, accommodate the expansion of water into ice and thus relieve the internal pressure generated, making the concrete highly resistant to freeze–thaw cycles.

To summarize, harsh environmental conditions may affect the mechanical properties of the FRP composite (degrading the fiber, the matrix, and the interface between them). Overall, moisture and alkaline environmental conditions have a more degrading influence on GFRP bars than the other environmental factors [123]. Table 3.4 shows the performance summary of FRP bars under harsh environmental conditions based on Benmokrane et al. criteria [162, 163]. The criteria are qualitative and depend only on the tensile strength degradation, being "excellent" if it is less than 10%, "good" if it is between 10% and 20%, "moderate" if it is between 20% and 30%, and "poor" if it is larger than 30%.

## 3.2 MECHANICAL EFFECTS

The most important time-dependent properties of FRP composites are creep (static fatigue), relaxation, and cyclic fatigue. However, other mechanical behaviors like shrinkage, deflection, and vibration could affect the serviceability performance of FRP-reinforced concrete elements.

**TABLE 3.4**

**Performance summary of FRP composites under harsh environmental conditions***

| Exposure/ Fiber | Glass | Basalt | Carbon | Aramid | Observations |
|---|---|---|---|---|---|
| Water exposure* | Excellent | Excellent | Excellent | – | Plasticization and hydrolysis cause cracks in the matrix resin. Swelling and dissolution degrade the fiber–matrix bond. Water ingress can also damage the fiber. Overall, fiber delamination and degradation of the FRP bar may cause tensile strength loss, reduction of both elastic modulus and interfacial shear strength. |
| Saline exposure* | Good | Good | Good | Good | Diffusion of NaCl or the presence of salts could strain or rupture intermolecular bonds in the matrix resin and the fiber–matrix interface. Composites with higher fiber volume fraction were found to absorb more seawater compared to ones with lower fiber volume fraction. Seawater solution caused greater moisture absorption than alkaline and tap water solutions. Degradation in tensile strength was observed. |
| Alkaline Exposure* | Good | Good | Excellent | Moderate | Alkaline solutions with pH values varying between 11.5 and 13.0 degrade the tensile strength and stiffness of FRP bars. The damage resulting from the presence of alkali is controlled by the matrix at the fiber–resin interface. The interlaminar-shear strength is the most affected property (with moderate degradation), followed by the transverse-shear and tensile strength, respectively. |
| UV light exposure | Good | Good | Good | – | The UV radiation may result in surface oxidation due to different chemical mechanisms related to resin type, capable of degrading the molecular bonds in the polymer matrix. FRP rebars embedded in concrete are themselves protected from the UV exposure but attention should be paid to storage. For external application, UV exposure can cause embrittlement and micro-cracking in an unprotected laminate surface. Color shift or yellowing and gloss changes are some of the indications of UV exposure on FRP laminates. |
| Elevated temperature** | – | – | – | – | Due to differences in coefficient of thermal expansion, high temperature variations can introduce FRP-reinforcement expansion causing splitting of surrounding concrete. Elevated temperature results into micro cracks at the interface between FRP and the concrete which decreases the FRP–substrate bond as well as the inter-laminar bond in the FRP. |
| Freeze–thaw cycles | Good | Good | Good | – | FRP materials become brittle under freezing temperatures and absorbed moisture expands when freezing leading to microcracks, particularly at the interface between the FRP and the substrate material. Exposure to freeze–thaw cycles could affect FRP composites themselves due to mismatched coefficients of thermal expansion of constituent resulting in micro-cracking and void generation. However, experimental studies on FRP bars have shown inconsistent behaviors when exposed to freeze–thaw cycles. |

*Performance depends on resins type, fiber volume fraction, solution concentration, aging temperature, among others. Further, elevated temperature accelerates degradation.

**Elevated temperature due to fire exposure can cause additional degradation. See Chapter 3 related to elevated temperature exposure.

### 3.2.1 Cyclic Fatigue

Cyclic fatigue, or commonly called fatigue, is a degradation of the capacity of a material caused by repeated applications of a large number of load cycles [46]. Most of the available data related to fatigue behavior of stand-alone FRP materials was obtained from aerospace applications [115]. Curtis [226] found that at one million cycles, the fatigue strength of carbon fibers is generally between 50% and 70% of initial static strength, with the resin playing an important role in the behavior. Results have indicated a continued downward trend (per decade of logarithmic life) of about 10%, 5%–6%, and 5%–8% for GFRP, AFRP, and CFRP, respectively when the ratio of the stress and the initial static strength is plotted versus the logarithm of the number of cycles at failure [52].

Odagiri et al. [183] reported a strength ranging from 54% to 73% of the initial bar strengths for two million cycle fatigue strengths of commercial AFRP bars. The failure was noted to be accelerated by the exposure to moisture and elevated temperature. Adimi et al. [227] observed that fatigue strength of CFRP bars encased in concrete decreased when the environmental temperature increased from 68 to 104°F (20 to 40°C). Fatigue tests on unbonded GFRP dowel bars have shown fatigue behavior similar to that of steel dowel bars for cyclic transverse shear loading of up to 10 million cycles. An equivalent performance can be achieved between FRP and steel bars subjected to transverse shear by changing some of the parameters, such as diameter, spacing, or both [228].

Studies on the effect of fatigue on the bond performance of deformed GFRP bars embedded in concrete have indicated that bond strength can increase, decrease, or remain the same following cyclic loading. This could be explained due to differences in surface treatments, materials, environments, and test methods [115]. Kumar and GangaRao [229] found no bond loss between GFRP bars and concrete after assessing the fatigue behavior of four full-size concrete bridge decks reinforced with sand-coated GFRP bars subjected to two million load cycles with a frequency of 1 Hz. The authors found the rate of degradation to be comparable to decks reinforced with steel.

El-Ragaby et al. [230] simulated vehicular loads by a stationary concentrated load varying cyclically in magnitude on GFRP-reinforced concrete deck slabs and observed that even after being subjected to fatigue loading of more than twice the expected lifetime of a real bridge, the specimens still had excellent performance in terms of load deflection response. The authors stated that the GFRP-RC bridge decks had a better fatigue performance and longer fatigue life (about three times) compared to those reinforced with steel.

According to ACI 440.2R-17 [5], CFRP systems perform well in fatigue under cyclic loading, while GFRP systems are more sensitive to this condition. The fatigue failure modes in CFRP and GFRP laminates have been identified as combinations of matrix cracking, transverse cracking, interfacial debonding, delamination, fiber breaking, etc. The failure is affected by several factors like the types of fibers and resins, textile architecture, surface treatments, fatigue stress ratios, and stress levels [231, 232]. A study carried out by Zhao et al. [231] showed that at low and medium fatigue stress levels, matrix cracking and interface debonding are the main failure modes for BFRP laminates.

Initial studies on RC beams strengthened with externally bonded CFRP laminates under fatigue loading showed failure modes that are controlled by fracture of the steel reinforcing bars in fatigue [117]. To corroborate these results, Breña et al. tested eight strengthened specimens under various stress ranges which found that fatigue did not affect the failure behavior (none failed due to fiber rupture, and bond fatigue was observed in only two specimens at high stress levels) but increased the deformations. Hence, the authors stated that the cyclic fatigue performance of RC elements strengthened with externally bonded FRP systems was comparable to that of RC elements. Ferrier et al. [233] observed a redistribution of stresses from the FRP laminate to the steel rebars during fatigue. In most cases, the failure has been observed to be initiated by successive yielding of the steel rebars in tension, leading to debonding of the CFRP laminate, the latter considered to be a secondary failure mode [234]. The same authors found that one layer of FRP laminate would increase the fatigue service load by 40% compared to a non-strengthened RC beam (to achieve this, rebar yielding in RC beams should not be permitted).

Zheng et al. [235] examined the fatigue performance of the bond between the carbon fiber laminate (CFL, pre-saturated laminates with epoxy adhesive) and concrete by carrying out 17 double shear tests at 60°C and 95% RH. The results showed a negative influence of the exposure on the bond performance and the fatigue life. The authors stated that the fatigue life (in terms of cycles of loading) may also be reduced by a higher stress level.

### 3.2.2 Creep Rupture

Static fatigue or creep rupture is a phenomenon produced by constant tension over time on FRP rebars that can suddenly fail after a period called the "endurance time" [115]. Thermoplastic matrix is not often applicable for structural engineering use on account of its low creep resistance. On the other hand, thermosetting matrix (polyesters, epoxies, and vinyl esters) has a higher creep resistance [236, 237]. Harsh environmental conditions and high sustained stress-to-strength ratio could shorten the endurance time [52, 238]. Moisture and fluid absorption decrease creep rate for concrete but increase creep rate in FRP bars, causing residual stress and degrading polymers, fibers, and fiber–matrix interfaces via hydrolysis and chemical attacks. Further, Singhvi and Mirmiran [150] stated that the presence of salt in the water solution does not affect the creep rate of FRP-RC in the same way as the moisture does. Another study by Al-Salloum and Almusallam [239] found that in GFRP-reinforced concrete beams exposed to environmental conditions at 40°C, the seawater wet–dry cycles were the harsher environment compared to tap water and seawater as it concerns the midspan deflection and strain. The susceptibility of the fiber to creep depends on the material, with carbon, aramid, and glass fibers being the least, moderately, and the most susceptible, respectively. Most polymeric resins can be susceptible to creep [46].

A linear relationship exists between creep rupture strength and the logarithm of time for times up to almost 100 hours obtained for a 6-mm diameter smooth GFRP, AFRP, and CFRP bar [240]. The ratio of stress at creep rupture after 50 years to the initial tensile strength was found to be about 30%, 50%, and 95% for GFRP, AFRP, and CFRP, respectively. Braided AFRP and twisted CFRP tendons (diameter varying between 5 and 15 mm and epoxy matrix) were found to have a ratio of 66% and 79%, respectively [238], 55% for a GFRP (E-glass/vinyelster) bar [241], and 85% at a projected 100 years for a 12.5 mm diameter CFRP twisted strand [115]. An investigation of creep deformation on CFRP and AFRP bars tested to 3000 hours reported a range of strains between 0.002% and 0.037% [134, 242]. The creep strain of CFRP, at room temperature and humidity, remains under 0.01% at a tensile stress of up to 80% of the tensile strength after 3000 h. Creep strain on GFRP bars was found to be less than 2.0% of the initial value after 4360 h under 27% of the ultimate tensile strength [216]. For AFRP and GFRP, the creep strain could range between 0.15–1% and 0.3–1.0%, respectively [46]. Table 3.5 shows the range of time-dependent creep parameters for some fiber types from available literature. More recently, Benmokrane et al. [243] collected and evaluated 204 creep-rupture tests from available literature and proposed a value of 50.7% of the average ultimate tensile strength for the creep-rupture strength at a 114-year endurance time for deformed, sand-coated, smooth, and deformed and sand-coated GFRP bars of sizes varying from 6 to 16 mm in diameter.

The FRP creep phenomenon is especially relevant for prestressed concrete elements. It was experimentally observed that creep rupture does not occur if sustained stress is lower than 60% of the short-term strength. In the RC elements, the low level of stress in FRP rebars at serviceability loads makes creep rupture unlikely [46]. The results of time-dependent bond slip tests on GFRP rebars due to sustained loading (15% of the GFRP bar tensile strength) showed an increase in the bond slip that stabilizes at approximately 60 days after loading, irrespective of the concrete strength. The concrete compressive strength influences the bond deterioration; the greater the concrete strength, the lower the creep slip [244].

### 3.2.3 Relaxation

Relaxation is a decrease in the load of the bar held with a prescribed initial load applied and held at a given constant strain at a given constant temperature [110]. Usually, relaxation is defined after one million hours. Test results indicate that at increasing temperature, the relaxation rate becomes greater, and this tendency is stronger for AFRP bars. Table 3.5 shows some ranges of relaxation for AFRP, CFRP, and GFRP after 1000 h and 50 years.

For the externally bonded FRP laminates, Choi et al. [245] carried out shear tests to investigate the time-dependent behavior of epoxy at the concrete–CFRP interfaces. The shear stress to shear strength ratio is a primary factor affecting the long-term behavior of the concrete–CFRP interfaces. The results indicated an ultimate creep coefficient (the ratio between the time-dependent creep strain and the instantaneous elastic strain) of 3.0 and a creep retardation time (the time when 63% of the

**TABLE 3.5**
**Range of time-dependent creep parameters [46]**

| Phenomenon | Aramid (%) | Carbon (%) | Glass (%) | Influencing Parameters |
|---|---|---|---|---|
| Creep strain under sustained load (i.e., 80% tensile strength after 3000 h) | 0.15–1.0 | < 0.1 | 0.3–1.0 | Temperature, humidity |
| Creep failure strength after about 50 years | 47–66 | 79–93 | 29–55 | Resin type, volume fraction, and orientation of fibers, environmental conditions |
| Relaxation 1000 h | 5.0–8.0 | 0.5–1.0 | 1.8–2.0 | Temperature, initial tensile stress |
| 50 years | 11–25 | 2.0–10 | 4.0–1.4 | |

creep has occurred) between 1 and 2 days compared to the typical retardation time of concrete that ranges between 300 and 900 days. Ferrier et al. [246] also evaluated the creep behavior of several epoxy adhesives on concrete–CFRP interfaces. The results showed that creep behavior is correlated with the glass transition temperature of the polymer. They recommended using polymer adhesives with high shear moduli (>10 GPa) and high glass transition temperatures (>55°C) to limit creep in the adhesive joint.

Al Chami et al. [247] tested CFRP-strengthened concrete beams subjected to high levels of sustained load varying from 59% to 78% of the ultimate static capacities of the unstrengthened beams for about one year. This was to account for the fact that strengthening is usually required when an element is expected to carry increased service loads. The results showed that the CFRP strengthening system increased the ultimate capacity of the beams, but there was no significant improvement in creep behavior.

### 3.2.4 Shrinkage

When manufacturing FRP composites, the resin could shrink during the curing cycle. This occurs due to a chemical loss of volume followed by thermal contraction during cool-down after cure [248]. Shrinkage volumetric reduction is in the range of 5–12%, 5–10%, and less than 5% for polyester, vinyl ester, and epoxy resins, respectively. This phenomenon could lead to porosity or micro-cracks [249]. The presence of voids and cracks is considered detrimental to the structural integrity of the composite and may cause reduced electrical resistance [250]. ASTM D5117 establishes a test method to observe the wicking action through delamination or longitudinal continuous voids. However, no specific values are indicated to imply an unsatisfactory performance of the composite.

Shrinkage in concrete itself is another phenomenon related to the time effect on the behavior of FRP-reinforced concrete structures, which may be assumed independent of load, but can also cause long-term deformations in concrete [244]. Plastic shrinkage cracking in concrete appears at its early stages of hardening when the rate at which the lost moisture is replaced by the bleed water is less than the rate of surface moisture evaporation [251].

Shao and Mirmiran [252] tested five different configurations of CFRP grids for the control of plastic shrinkage cracking of concrete using CFRP rods. Test results showed that carbon FRP grids meet or exceed the acceptance criteria (according to ICC-ES [253] for synthetic fibers) for providing a minimum 40% reduction in shrinkage cracking for concrete. However, there is still a lack of information and guidelines related to shrinkage and secondary reinforcement configurations that use FRP material.

## 3.3 COMPARISON BETWEEN SERVICEABILITY CRITERIA FOR CONVENTIONAL STEEL-REINFORCED CONCRETE AND FRP-RC

FRP-reinforced concrete members have a relatively smaller stiffness after cracking when compared to steel-reinforced concrete members of identical size and reinforcement layout [115]. Because of the variable stiffness, brittle-elastic nature, and particular bond features of FRP reinforcement, FRP-RC members are more sensitive to the variables affecting deflection. Generally, FRP design for flexural strength may not satisfy serviceability criteria for deflection and crack control [254–256], and consequently, serviceability limit states (deflection, crack control, creep, fatigue) can control the design of FRP-RC members. When compared to steel-reinforced concrete, an equal area of FRP will result in larger deflections and wider cracks in the FRP-RC element. The neutral axis depth for the balanced section in the FRP-RC member is close to the compressive end. Additionally, higher compressive strains in the concrete are expected in the FRP-reinforced section for the same beam depth and applied moment as the steel-reinforced section [257].

However, the serviceability criteria for crack width and deflections are generally satisfied when a section is designed to achieve concrete crushing failure prior to tensile rupture of the FRP rebar [52]. At present, most of the available design guides and recommendations for FRP-reinforced concrete elements are focused on GFRP rebars [110, 111, 114, 185] and CFRP strands [258]. Given the brittle behavior of both FRP reinforcement and concrete, compression- and tension-controlled sections are acceptable in the design of flexural members reinforced with FRP bars, with a strength reduction factor ranging from 0.75 to 0.55, respectively [111]. This is different from steel-RC design, where the compression-controlled sections have a lower strength reduction factor (0.65–0.75) compared to tension-controlled sections (0.90). Table 3.6 shows the strength reduction factors related to GFRP-RC according to different design guides and codes for FRP- and steel-reinforced concrete elements.

The strength and stiffness of the FRP composite may change under harsh environmental conditions that could limit the service life of the structure and, although

**TABLE 3.6**
**Strength reduction factors related to GFRP-RC for different design guides and codes**

| | GFRP | | Steel | |
|---|---|---|---|---|
| **Action** | **ACI 440 (2022)** | **AASHTO GFRP (2018a)** | **ACI 318-19 (2019)** | **AASHTO LRFD (2020)** |
| Moment, axial force, or combined moment and axial force | 0.65–0.55* | 0.75–0.55* | 0.65–0.90 | 0.75–0.90 |
| Shear | 0.75 | 0.75 | 0.75 | 0.85 |
| Torsion | 0.75 | 0.75 | 0.75 | 0.85 |
| Bearing | 0.65 | 0.70 | 0.65 | 0.70 |

*Factors listed from compression to tension controlled.

FRP rebars do not exhibit corrosion the same way as steel rebars, many studies, as referenced in the previous chapter, have reported deteriorations that result in tensile and bond strength loss when exposed to harsh environments such as moisture and alkaline solutions. However, unlike the visible warning signs of steel-RC elements (i.e., cracking or spalling of the concrete), the deterioration of FRP rebar does not result in a (noticeable) increase in volume and hence does not lead to cracking and/or spalling of the concrete cover [130, 261]. Therefore, the durability requirements for FRP-RC structures are different from those utilizing steel-RC because of the corrosion-resistance properties of FRP bars. Consequently, some of the design criteria that aim at mitigating the corrosion of the rebar such as increased concrete cover, the use of corrosion-inhibiting admixtures, the use of epoxy coatings, and limitations on crack widths to delay the initiation of corrosion are not necessarily applicable to or necessary for FRP-RC structures [114]. Unlike steel–RC structural members, for FRP–RC members, there is no special concrete cover thickness provision required for corrosion control [196].

The absence of a warning for failure of the FRP-RC element when exposed to harsh environmental conditions can be seen as a deficiency for the FRP-RC members because of their brittle failure characteristic [121]. Consequently, design guides for FRP-RC establish the design tensile strength of FRP bars as the guaranteed tensile strength (value reported by the manufacturer computed as no larger than the mean minus three standard deviations) times an environmental reduction factor to account for the long-term effects of environmental exposure. Table 3.7 presents the environmental reduction factors on FRP-reinforced concrete structures for different design guides.

AASHTO [111] states that durability conditions related to concrete remain the same as those addressed in steel-reinforced concrete elements, and they could be found in AASHTO LRFD Bridge Design Specifications [260]. The new ACI GFRP

**TABLE 3.7**
**Environmental reduction factors on FRP-reinforced concrete structures**

| FRP Composite | ACI 440 (2022) $C_E$ | AASHTO GFRP Specifications (2018a) | CSA S806 [185] | JPCI [110] |
|---|---|---|---|---|
| Glass | 0.85 | 0.7–0.8 | *Not specified* | *Not specified* |
| Carbon | 0.9–1.0* | *Not specified* | | |
| Basalt | *Not specified* | | | |
| Aramid | 0.8–0.9* | | | |

*No longer applicable, used in the previous ACI 440.1R-15 design guide.

code [114] also establishes requirements for concrete mixtures in Chapters 19.3.2 and 26.4, considering applicable environmental exposure to provide the required durability.

Another critical serviceability limit state is fire resistance. The critical glass transition temperature of FRP is, in general, much lower than that causing loss of stiffness in steel, and the impact of elevated temperatures on the behavior of FRP composites is rapid and causes severe degradation of both cross-sectional and bond properties. The critical temperature for tensile reinforcement has historically been defined as the temperature at which it loses much of its strength (50% loss of room temperature yield strength) and can no longer support the applied load. For reinforcing steel, the critical temperature is 593°C [262]. However, for FRP composites, there is no specific critical temperature value. AASHTO [111] specifies a minimum concrete cover for GFRP bars ranging from 1.0 to 2.0 times bar diameters for all exposure conditions except for additional fire protection. However, no additional recommendation on the latter has been established. As mentioned earlier for the effect of the elevated temperature exposure, if GFRP bars are well anchored outside of the area directly exposed to fire, they can retain considerable strength and stiffness during a fire event [199, 200]. ACI 440 [114] also establishes minimum concrete cover for cast-in-place and precast concrete members in its Table 20.6.1.3.1 with a complementary table in the comments section with fire resistance rating provided by the minimum covers.

To avoid creep rupture of GFRP reinforcing bars, the sustained stress level shall be limited by the creep rupture stress. Because these stress levels will be within the elastic range of the member, the stresses can be computed through an elastic analysis [111]. Using the full live load in determining the applied stress is prohibitively conservative. Therefore, the load factor applied to live load is reduced from 1.0 to 0.2 to account for the sustained portion of the live load [10]. For common steel RC, no additional long-term effects need to be considered because steel undergoes neither shrinkage nor creep [244]. The maximum sustained tensile stress in the FRP reinforcement shall be less than the design tensile strength of FRP reinforcing bars times a creep rupture reduction factor (equal to 0.30 following ASTM D7337/D7337M)

**TABLE 3.8**

**Creep rupture reduction factor for tensile strength in different design guides and codes**

| | FRP | | | Steel | | |
|---|---|---|---|---|---|---|
| **Rebar** | **ACI 440 (2022)** | **AASHTO GFRP (2018b)** | **CSA S806* [185]** | **Rebar** | **ACI 318 [259]** | **AASHTO Bridge Design (2020)** |
| GFRP | 0.30 | 0.30 | 0.25 | Steel | No reduction | No reduction |
| CFRP | 0.55** | – | 0.65 | | | |
| BFRP | *Not specified* | – | – | | | |
| AFRP | 0.30** | – | 0.35 | | | |

*Multiplied by the guaranteed tensile strength and not the design tensile strength.
**No longer applicable.

[111]. Table 3.8 shows the creep rupture reduction factor for different design guides and codes, for FRP and steel design.

The use of GFRP bars as compression reinforcement of flexural members is not recommended. However, placing GFRP rebars in the compression zone of flexural members is permitted, provided that they are not considered for the determination of the member flexural resistance [111]. This is because the limited compressive strength and modulus of the FRP bars do not increase the strength nor reduce the effect of concrete creep of FRP-reinforced flexural members [10].

The limit for maximum crack width of FRP-reinforced concrete by CSA S6:19 [263] is 0.020 in. (0.50 mm) for members subjected to aggressive environments and 0.028 in. (0.70 mm) for other members. However, there is no such provision of crack control in ACI 440 [115] for FRP-reinforced concrete (the maximum crack width for steel reinforcement varies between 0.018 and 0.022 in. (0.46 and 0.56 mm)). The ACI 440 [114] code provisions for reinforcement spacing are intended to limit surface cracks to a width that is generally acceptable in practice but may vary widely in a given structure. Large flexural cracks may cause public discomfort due to a perceived risk of structural collapse, even though they may be structurally safe. The crack width limits for FRP-reinforced concrete elements under service are recommended to limit fluid ingress that could degrade the FRP bars while also ensuring their acceptable aesthetic appearance [178].

# 4 Damages and Defects in FRP Applications

## 4.1 DAMAGES AND DEFECTS IN FRP STRENGTHENED CONCRETE ELEMENTS (EXTERNAL APPLICATION)

This section will discuss the common defects for external FRP applications, and the following section discusses the defects associated with internal applications.

The deficiencies in the external application of FRP are commonly categorized based on their occurrence within different material interfaces and components. These include: (A) within the FRP composite itself, (B–D) at the bond interfaces (e.g., FRP-adhesive, adhesive, adhesive-concrete), and (E–F) within the concrete substrate or the reinforcement interface, as shown in Figure 4.1. To clarify the origin and nature of these deficiencies, Table 4.1 systematically classifies them into three main categories: FRP composite defects, bond defects, and defects in concrete. Each category is further divided by specific defect locations and the associated defect types commonly observed in practice. This structured format enables a clearer understanding of the mechanisms leading to performance degradation in externally strengthened FRP bridge elements.

### 4.1.1 FRP Composite Deficiencies

*A.1 Blisters*

Blisters are bubble-like formations that occur on the surfaces of externally applied FRP composites, as shown in Figure 4.2. These appear due to the combined action of freeze-thaw cycles and entrapped moisture at the contact between the substrate and the FRP material or at the inter-laminar interfaces of the composites. Nevertheless, because their effects are primarily restricted to the surface, this imperfection has little impact on the structural performance of the structure.

*A.2 Wrinkling*

Wrinkling appears as creases or folds on the surface of the FRP composites (Figure 4.3), often occurring at corners and curves of the structure. It is caused by improper installation practices during the hand lay-up process, such as excessive stretching or shearing of the fabric. The safety of the structure may be compromised if it results in insufficient surface contact of the FRP composites with the substrate.


DOI: 10.1201/9781003646969-4

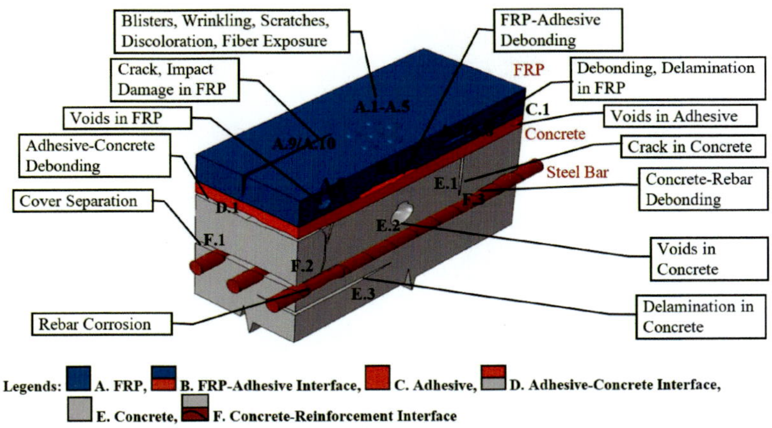


**FIGURE 4.1** Deficiencies in external FRP application

## TABLE 4.1
## Classification of deficiencies in external FRP application

| Defect Categories | Defect Locations | Defects |
|---|---|---|
| FRP composite defects | A. FRP | A.1-A.5. Surface defects (A.1 Blisters, A.2 Wrinkling, A.3 Scratches, A.4 Discoloration, A.5 Fiber Exposure) |
| | | A.6. Voids in FRP |
| | | A.7/A.8. Debonding/Delamination in FRP |
| | | A.9/A.10. Cracks/Impact damage in FRP |
| Bond defects | B. FRP–adhesive interface | B.1. FRP–adhesive debonding |
| | C. Adhesive | C.1. Voids in adhesive |
| | D. Adhesive–concrete interface | D.1. Adhesive–concrete debonding |
| Defects in concrete | E. Concrete | E.1 Cracks in concrete |
| | | E.2. Voids in concrete |
| | | E.3 Delamination/spalling in concrete |
| | F. Concrete–reinforcement interface/ the plane passing through the reinforcement layer | F.1 Cover separation |
| | | F.2 Corrosion in steel reinforcement |
| | | F.3 Concrete–reinforcement debonding |

*A.3 Scratches*

Scratches at the surface of externally applied FRP range from superficial scratches to deeper grooves (Figure 4.4). They can occur anytime during the installation and service life of the structure. They are detrimental if they develop into full-depth cracks as they propagate deep into the fibers of the composite.

**FIGURE 4.2** Blistering in FRP

**FIGURE 4.3** Wrinkling

*A.4 Discoloration*
Discoloration (Figure 4.5) is mainly caused by exposure to UV rays, heat, chemicals, fire, excessive strain, subsurface defects, voids, and moisture penetration. It may be a sign of composite degradation, which may be followed by cracks and embrittlement.

**FIGURE 4.4** Scratch on the externally applied FRP ( Source: FHWA)

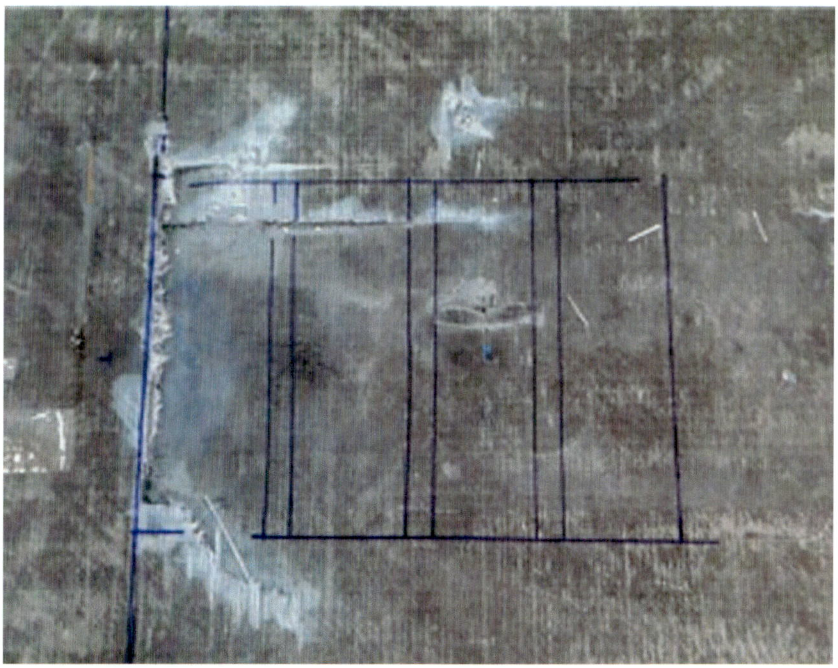

**FIGURE 4.5** Discoloration [28]

*A.5 Fiber Exposure*

Fiber exposure, as shown in Figure 4.6, are simply exposed fibers of the FRP composite caused mainly by improper handling and installation of FRP composites.

**FIGURE 4.6** Fiber exposure [28]

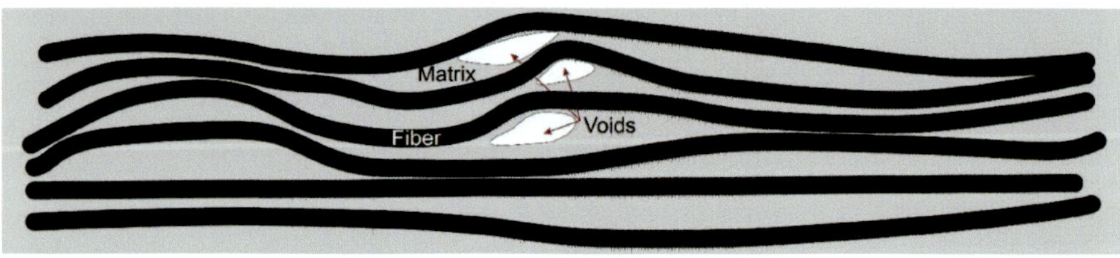


**FIGURE 4.7** Voids inside composite material at the interlaminar interfaces

These are the entry points of moisture and contamination into the composite, which deteriorate the composite properties.

*A.6 Voids: Fiber–Matrix Interface*

Voids present at the fiber–matrix interface, as shown in Figure 4.7, are the cavities formed due to entrapped air within the layers of the composites. These are caused by the overlapping of the fabrics during fabrication or installation. FRP composites with voids are the result of bad workmanship, which reduces their laminar shear strength. The fabric/matrix defects are only related to pre-cured laminates externally bonded to concrete or FRP bars for NSM (near surface mounted) installation. In these cases, the fiber/matrix defects can and should be identified by the QC (quality control) process of the manufacturer.

*A.7 Fiber–Matrix Debonding*

Debonding within FRP composites is the separation at the interface between the fiber and the matrix, as shown in Figure 4.8. It is mainly caused by the presence of

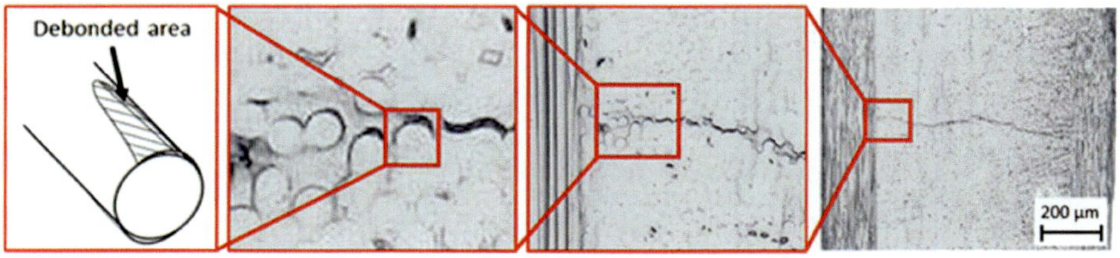


**FIGURE 4.8** Debonding between fiber and matrix [264]

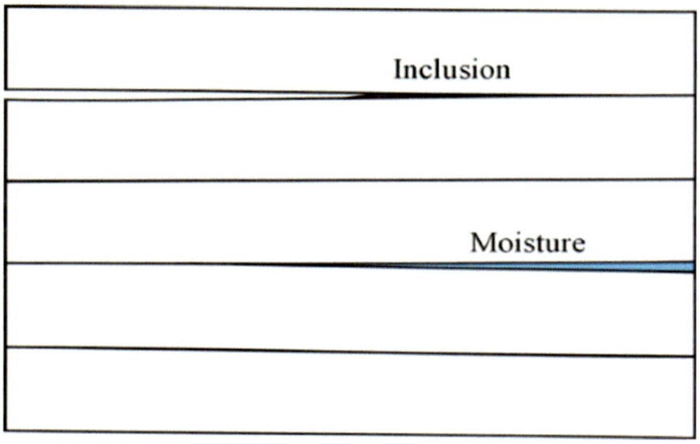


**FIGURE 4.9** Delamination in FRP laminate [265]

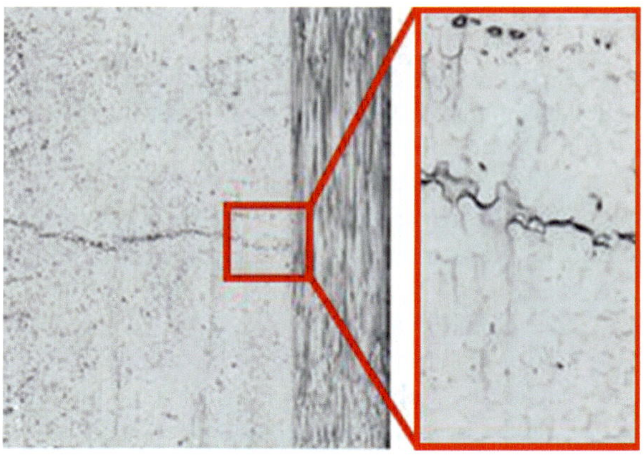

**FIGURE 4.10** Crack formation in FRP composite [264]

surface moisture on the fibers. The effects of debonding include a loss of composite action, which reduces the material's ability to provide strength against transverse tension, inter-laminar shear, and impact.

*A.8 Delamination between Composite Layers*

Delamination within FRP composites is the separation at the interface between the layers of FRP laminae comprising the composite, as shown in Figure 4.9. It is often

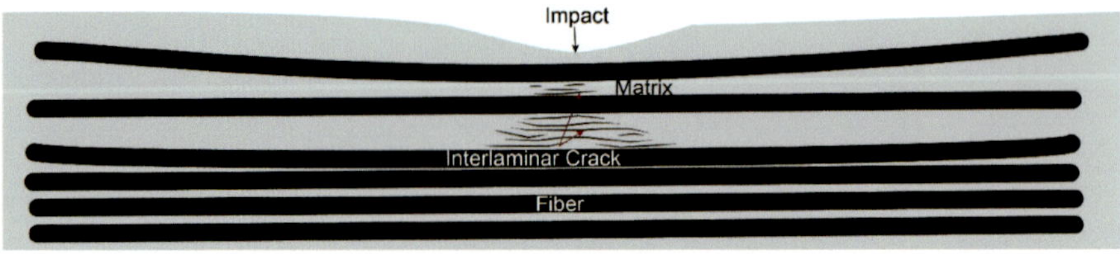


**FIGURE 4.11** Damage at surface and subsurface due to impact

caused by moisture, foreign object contamination, and trapped air between the FRP layers. Delamination can lead to a loss of shear transfer capacity in the material.

*A.9 Cracks*

Cracks in FRP composites (Figure 4.10) mostly occur in the direction parallel to the fiber. They are caused by entrapped air, inadequate resin distribution, insufficient reinforcing fibers, impact, and service loads. Cracks may lead to failure if they propagate deeper and expand under loading conditions.

*A.10 Impact Damage*

Impact damage (Figure 4.11) is caused by moving objects. Slow-moving objects cause damage that is not critical at the surface but significant at the internal level. Fast-moving objects cause damage at the surface level.

### 4.1.2 Bond Deficiencies

*B.1/D.1 Debonding between FRP and Concrete*

Debonding (Figure 4.12) is the loss of cohesive and adhesive bond between the externally applied FRP and the concrete substrate. It can occur when flexural or shear cracks open under high loading conditions. It can be caused by improper installation practices, insufficient curing of resin, inadequate surface preparation, or surface moisture on the substrate. Excessive debonding may lead to brittle fracture of the concrete element underneath the external FRP as the composite fails to transfer the stresses to the substrate.

*B.2/C.1/D.2 Voids between FRP and Concrete*

Voids (Figure 4.13) are areas of zero contact between the FRP composites and the concrete substrate. They are typically formed due to air trapped in between the layers of composites, air mixed or volatiles contaminated, or insufficient resin matrix,

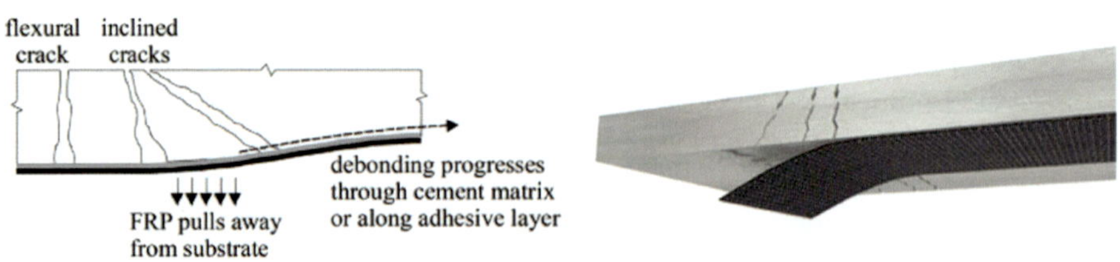


**FIGURE 4.12** Debonding of externally bonded FRP [5]

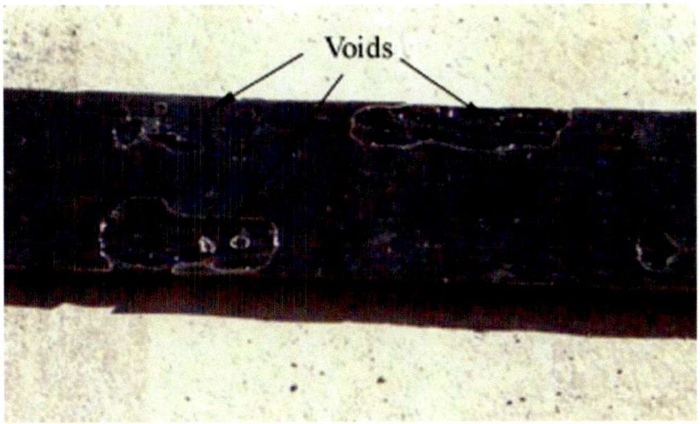


**FIGURE 4.13** Voids at FRP–concrete interface [265]

and irregular surface of the substrate. Voids can also form "bubbles" when moisture from the substrate evaporates and cannot escape prior to complete polymerization of the resin. Since voids cause regions of stress concentrations, they weaken the bond strength of the FRP application.

### 4.1.3 Deficiencies in Concrete

*E.1 Cracks*

Cracks in the concrete substrate are completely hidden behind the externally applied FRP. They are caused by shrinkage, thermal stresses, chemical exposure, weathering, corrosion of steel bars, design errors, poor detailing and construction practices, and excessive loading conditions. Cracks are the points of entry for detrimental chemicals that attack steel reinforcement. Their presence in externally applied concrete elements weakens the bond between FRP and concrete (Figure 4.14).

*E.2 Voids in Concrete*

The formation of voids in concrete is independent of the external application of FRP and related to the inadequate design and construction measures during the casting

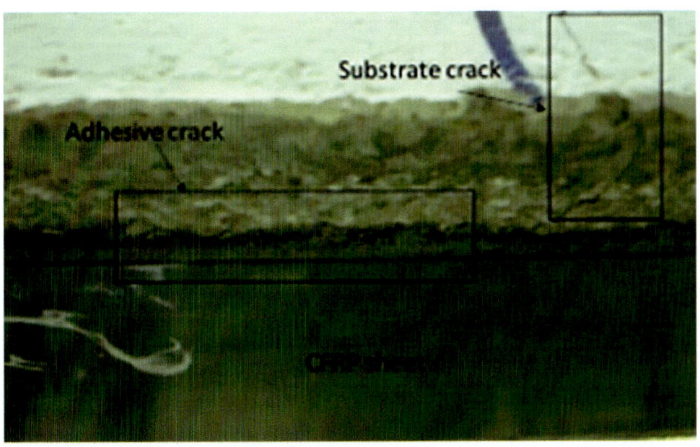


**FIGURE 4.14** Cracks in concrete substrate [266]

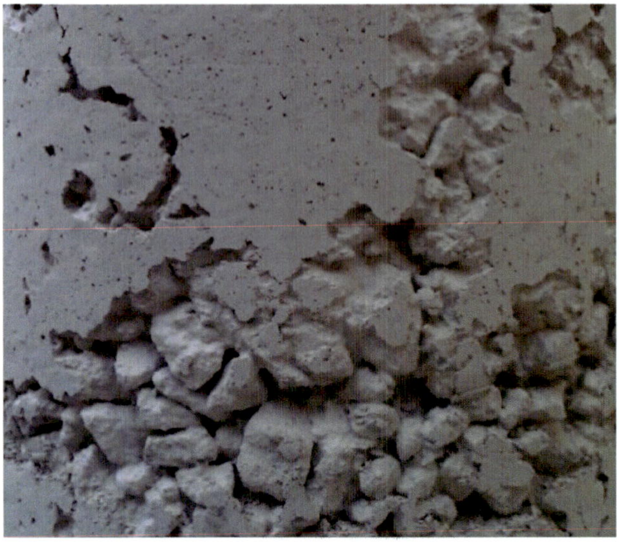

**FIGURE 4.15** Concrete voids

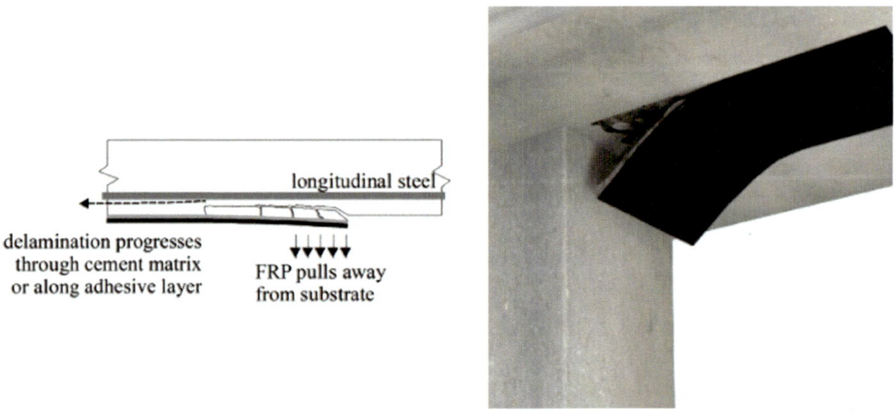


**FIGURE 4.16** Concrete delamination [5]

of the concrete substrate. Voids can form due to improper vibration of concrete during casting, in addition to stiff or unworkable concrete, segregation, crowded rebar, insufficient concrete consolidation, and irregular aggregate sizing. It may cause concrete structures to deteriorate over time. Figure 4.15 shows voids formed on the surface of concrete.

*E.3 Concrete Delamination*

Concrete substrate is weaker than the adhesive and the FRP used in an external application. Hence, delamination of concrete (Figure 4.16) occurs when the stresses and forces in FRP are high enough to rip out the concrete beneath it. It initiates near cracks or termination points of external FRP systems where the interfacial shear and normal stresses build up excessively. Delamination failures are sudden and brittle.

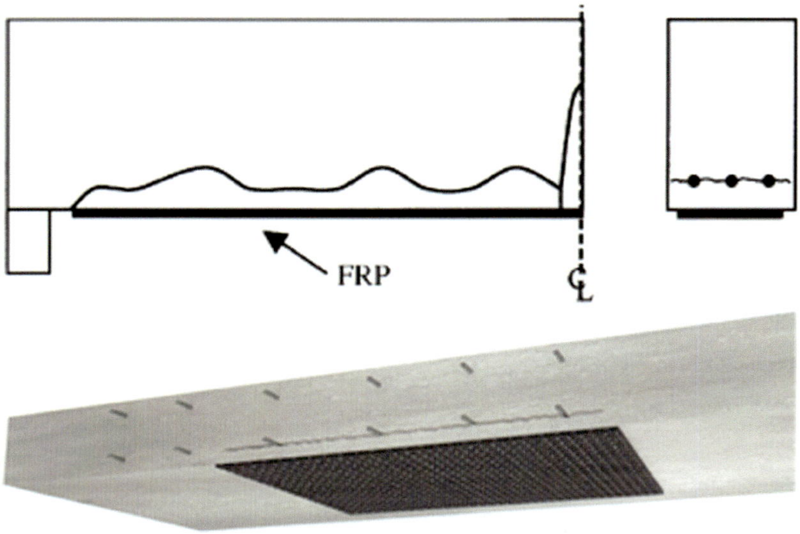


**FIGURE 4.17** Cover separation [5]

**FIGURE 4.18** *Corrosion in steel reinforcement embedded in concrete substrate* [267]

*F.1 Cover Separation*

While delamination occurs just beneath the externally applied FRP, cover separation (Figure 4.17) occurs at a deeper level up to the cover distance of internal reinforcement, which provides a weaker horizontal plane for the separation to progress. Cover separation is also often referred to as FRP end peeling, concrete cover delamination, end-of-plate failure through concrete, concrete rip-off failure, debonding at rebar layer, and local shear failure. It occurs as the cracks that have reached up to the internal reinforcement begin to propagate horizontally along the level of reinforcement under high stresses in the presence of externally bonded FRP. Similar to delamination, it is a sudden and brittle failure.

*F.2 Corrosion in Steel Reinforcement Embedded in Concrete Substrate*

External FRP is typically applied to strengthen steel-reinforced concrete elements. The corrosion of steel reinforcement will not be completely stopped, even though external FRP might decrease the rate of corrosion (Figure 4.18). Corrosion activity should be monitored on concrete elements even after strengthening measures have been applied.

**TABLE 4.2**
**Summary of defects in external applications of FRP**

| When \ Where | Defects Intrinsic to the FRP Composites | Bond Defects | Defects in Concrete |
|---|---|---|---|
| *Manufacturing and installation/ casting defects* | Voids at fiber–matrix interface, wrinkling, blistering, fiber–matrix debonding, delamination between composite layers, fiber exposure, scratches, cracks, discoloration | Voids at FRP–concrete interface, debonding between FRP and concrete | Voids, cracks |
| In-service defects | Voids at fiber–matrix interface, fiber–matrix debonding, delamination between composite layers, fiber exposure, scratches, cracks, discoloration, impact damage | Debonding between FRP and concrete | Delamination/ cover separation, cracks, corrosion in steel reinforcement |

### 4.1.4 Summary of Defects in External Applications of FRP

The defects in external FRP applications are categorized into three groups based on their location: defects intrinsic to the FRP composites, bond defects, and defects in concrete. In addition to the classification based on where they occur, the defects can also be divided into manufacturing and installation/casting defects and in-service defects based on when the defects are initiated. The summary of defects in external applications of FRP based on where and when these defects occur is shown in Table 4.2.

## 4.2 POTENTIAL DAMAGES AND DEFECTS IN FRP-REINFORCED CONCRETE ELEMENTS (INTERNAL APPLICATION)

In the case of internal application of FRP, the defects are possible at: G. FRP reinforcement, H. Concrete–FRP interface, and I. Concrete, as shown in Figure 4.19 and Table 4.3. They are classified as: defects in FRP reinforcement, bond defects, and defects in concrete.

### 4.2.1 Defects in FRP Reinforcement

Defects such as voids at fiber–matrix interface, fiber–matrix debonding, delamination between composite layers, fiber exposure, scratches, cracks, and discoloration can also occur in FRP rebars during manufacturing, transportation, storage and handling, and service.

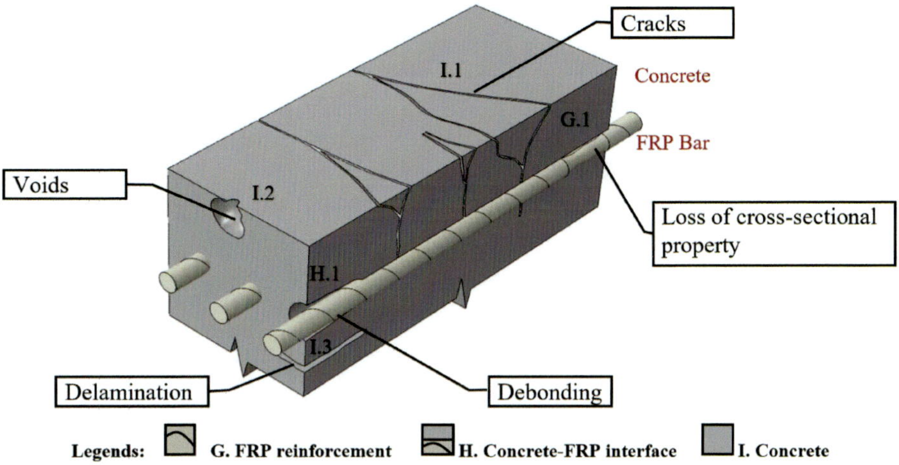


**FIGURE 4.19** Potential deficiencies in internal FRP application

**TABLE 4.3**
**Classification of potential deficiencies in internal FRP application**

| Defect Categories | Defect Locations | Defects |
|---|---|---|
| Defects in FRP reinforcement | G. FRP reinforcement | G.1. Loss of cross-sectional property<br>(other potential defects: voids at fiber–matrix interface, fiber–matrix debonding, delamination between composite layers, fiber exposure, scratches, cracks, discoloration) |
| Bond defects | H. Concrete–FRP reinforcement interface | H.1 Debonding<br>(other potential defects: H.2 slippage) |
| Defects in concrete | I. Concrete | I.1 Cracks<br>I.2 Voids<br>I.3 Delamination |

*G.1 Loss of Cross-Sectional Properties*

The service life of FRP-reinforced concrete members can be reduced due to chemical attacks of alkaline, saline, and other extreme environmental conditions. Degradation of FRP reinforcement due to chemical effects is a result of the corresponding degradation of the fibers and the matrix, as well as the fiber–matrix interface. There might not be visible or physical evidence of chemical attacks on FRP rebars embedded in concrete despite the reduction in their mechanical properties, which hinders the detection of the loss of cross-sectional physical and mechanical properties.

The available data mainly focuses on GFRP and BFRP bars for internal use, as well as CFRP laminates for external applications. However, due to variations in parameters and a lack of standardization, establishing clear degradation patterns, even for specific composites like GFRP/vinyl ester bars, can be challenging.

Improvements in material properties over time also make earlier data less representative of modern FRP composites. Notably, alkaline exposure is identified as the most damaging condition, with CFRP displaying superior performance in aggressive environments, while GFRP exhibits resilience across various conditions. The environmental reduction factor available in the U.S. design guidelines accounts for the high pH level of both pore-water solutions and the presence of alkali ions, the mean temperature, and the humidity, for an assumed service life of 75–100 years [163].

### 4.2.2 Bond Defects

*H.1 Debonding*

The structural performance of RC members primarily depends on the adequate bond between the concrete and the reinforcing bars. The bond ensures that the high tensile strength of the FRP rebars is used effectively. The tensile and the bond properties of the FRP rebars are of key importance for internal application in RC members. However, with the passage of time, the bond could gradually deteriorate due to environmental and load effects, potentially resulting in bond failure of the FRP-RC members compromising the integrity of the structure.

*H.2 Slippage*

Slippage, a relative displacement between the reinforcing bars and the concrete, occurs when external forces overcome the strength of the bond between them. Monitoring bond-slip is important because the structural integrity and overall strength of concrete structures depend on the condition and strength of the bond between the reinforcement and the concrete. Hence, more attention should be given to the detection and prediction of the occurrence of bond-slip to issue early warnings.

### 4.2.3 Defects in Concrete

The defects in concrete, such as I.2 voids and I.3 delamination as described in external application (E.2, E.3), are the same whether FRP is used for strengthening or reinforcing purposes. However, the cracks in FRP-RC are discussed separately under this section, as their formation is mostly associated with the difference in mechanical properties of FRP (low modulus of elasticity in FRP reinforcements) compared to steel.

*I.1 Cracks*

Cracks are the predominant defect reported by the bridge inspectors in the inspection of FRP-RC bridge decks, as shown in Figures 4.20 and 4.21, which can be attributed to the low modulus of elasticity of the FRP rebars. The initiation of cracks and large deflections in the concrete members reinforced with FRP rebars are one of the potential defects to be detected in the internal application of FRP composites, which raise serviceability concerns (see Figure 4.22).

FIGURE 4.20 Top surface cracks in GFRP-reinforced bridge deck [268]

FIGURE 4.21 Full depth crack in GFRP-reinforced bridge deck [268]

### 4.2.4 Summary of Defects in Internal Applications of FRP

The summary of defects in internal applications of FRP based on where and when these defects occur is shown in Table 4.4.

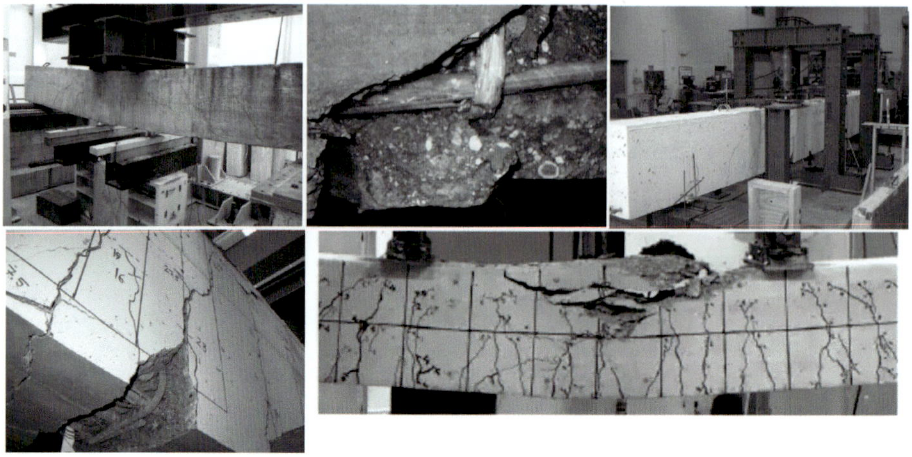

FIGURE 4.22 Cracks and failure modes in FRP-reinforced concrete members [10, 269]

## TABLE 4.4
## Summary of defects in internal applications of FRP

| When \ Where | Defects Intrinsic to FRP Composites | Defects in the Interface | Defects in Concrete |
|---|---|---|---|
| Manufacturing and construction defects | Voids at fiber–matrix interface, wrinkling, blistering, fiber–matrix debonding, delamination between composite layers, fiber exposure, scratches, cracks, discoloration | | Voids, cracks |
| In-service defects | Voids at fiber–matrix interface, fiber–matrix debonding, delamination between composite layers, fiber exposure, scratches, cracks, loss of cross-sectional property | Debonding, slippage, anchorage failure | Cracks |

## 4.3 SUMMARY OF DEFECTS IN FRP APPLICATION

All the defects identified in the previous sections of this chapter for both internal and external applications of FRP have been summarized in a comprehensive flowchart as shown in Figure 4.23.

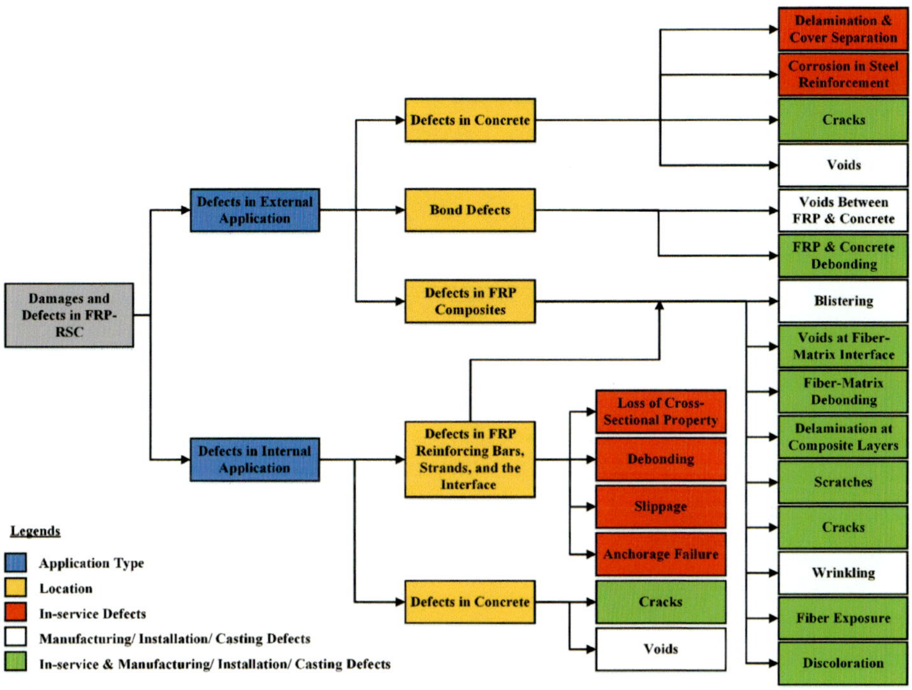


**FIGURE 4.23** Damages and defects in FRP application

## 4.4 POTENTIAL SOURCES OF DAMAGES TO LOOK FOR

The sources of deficiencies related to FRP composites and concrete are broadly divided into design factors, mechanical factors, environmental factors, and fabrication and workmanship errors. The presence of these sources would imply that deficiencies could be expected. A rational relationship between observed or presumed defects in the FRP-RSC elements, sequence of occurrence, and their causes need to be established and analyzed as defect etiology. A reliable etiology that takes into account the specific characteristics of such elements is believed to be essential for effective and accurate inspection and damage detection, and their impact on serviceability and structural performance. The sources of deficiencies in FRP composites and the corresponding bond/interface defects shown in Figure 4.24 are covered in the following sections (Sections 4.4.1 to 4.4.4). However, for the source of deficiencies in concrete shown in Figure 4.25, a study by Mehrabi and Farhangdoust [270] on the NDT methods applicable to health monitoring of accelerated bridge construction (ABC) closure joints can be consulted.

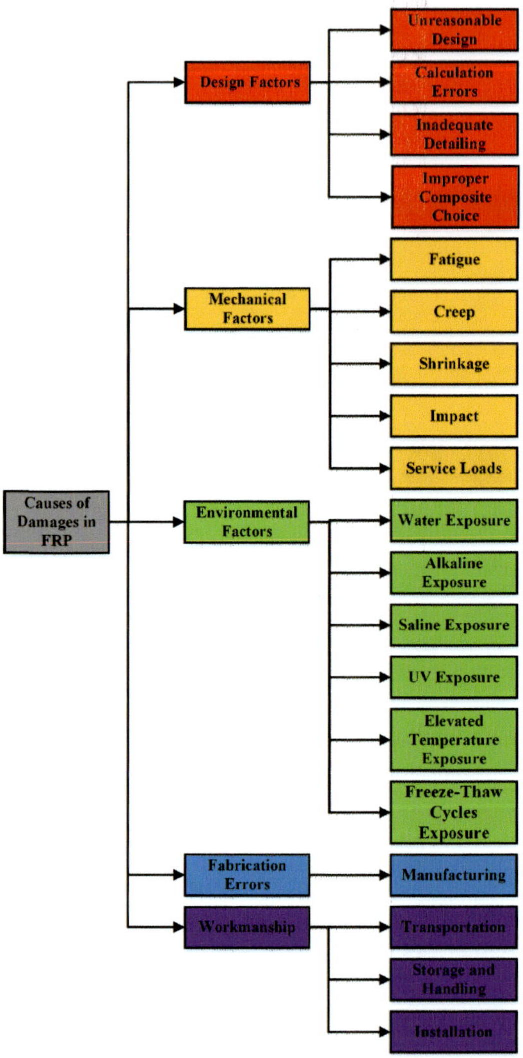


**FIGURE 4.24** Sources of deficiency in FRP

### 4.4.1 Design Factors

The design factors causing defects in FRP application can be referred to as the deficiencies in the design process (can be a lack of appropriate design specifications and codes, calculation errors, mistakes in following specifications, issues with constructability, improper engineering decisions, and improper composite choice). Design errors can be avoided by rigorous checks of the design prior to the commencement of construction and ensuring performance-based design taking fatigue, possible

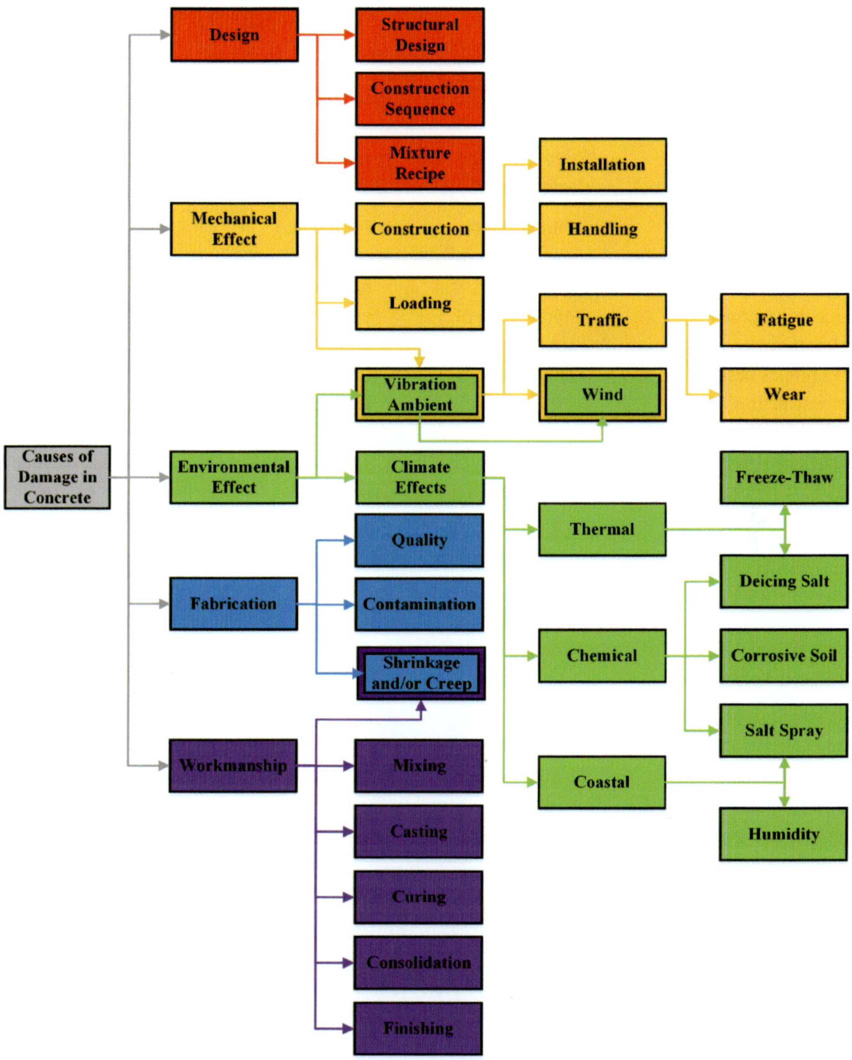


**FIGURE 4.25** Sources of deficiency in concrete [270]

overload, and durability into consideration. The design errors can be mainly categorized as shown in Table 4.5.

### 4.4.2 Mechanical Factors

The FRP-RSC elements are subjected to dynamic (impact), cyclic (fatigue), and static (creep) mechanical loading conditions along with shrinkage and service loads. These mechanical factors could potentially deteriorate the mechanical properties

**TABLE 4.5**
**Design factors causing deficiencies in FRP**

| Unreasonable design | Calculation or design errors | Inadequate installation or construction details | Improper composite choice |
|---|---|---|---|
| • Inappropriate decisions taken for the application of rehabilitation and repair systems<br>• Improper design for construction using FRP reinforcements<br>• Lack of appropriate design specifications and codes | • Human errors in calculation<br>• Improper application of design specifications | • Inadequate detailing for the installation of external and construction with internal FRP<br>• Lack of constructability checks | • Improper selection of different parameters such as fiber and matrix types, fiber length and distribution, and resin content |

of the FRP system either gradually over time or instantly, which has been further discussed below.

#### 4.4.2.1 Cyclic Fatigue

Fatigue is the degradation in the mechanical properties of a material under repeated loads. The behavior of FRP composites under fatigue is different from that of metals. In metals, there are two distinct phases—damage initiation and damage propagation—of a single crack under fatigue. Whereas, in FRP composites, the controlling mechanisms follow an accumulation of damage in a global fashion without the initiation phase rather than in a localized fashion as it is in metals. These include fiber-matrix debonding, matrix cracking, delamination, and fiber fracture. The damage mechanisms in externally applied FRP composites under fatigue are matrix cracking, transverse cracking, interfacial debonding, delamination, and fiber breaking. For internal FRP reinforcement, fatigue can lead to deterioration of the bond strength of the embedded FRP bars depending on the surface treatments, materials, environment, and loading conditions.

#### 4.4.2.2 Creep Rupture

FRP composites may undergo creep rupture when micro-cracks and voids introduced during the FRP fabrication process further develop under sustained loading conditions and are exposed to moist, saline, or alkaline environments, as shown in Figure 4.26. Creep can also lead to bond slip in internal FRP rebars through the mechanism of volume change; however, with the use of concrete with higher compressive strength, the creep slip can be lowered.

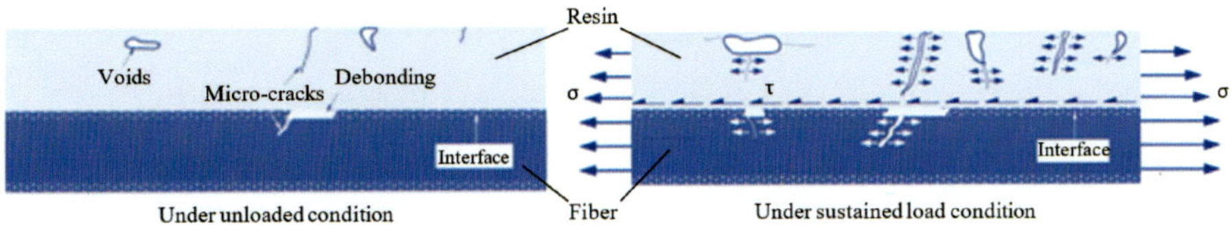


FIGURE 4.26 Deterioration under sustained load (Adapted from: [271])

#### 4.4.2.3 Unrestrained Shrinkage

FRP composites shrink due to a decrease in the volume of resin due to chemical loss as well as by thermal contraction during the curing cycle. This could result in the formation of voids and cracks in the FRP composite. However, it should be noted that defects due to shrinkage in FRP composite can only occur during manufacturing processes.

#### 4.4.2.4 Impact

Impact loads due to moving objects might cause either clearly visible surface defects or hidden subsurface defects in FRP-RSC elements.

### 4.4.3 Environmental Factors

The extended exposure of FRP composites to different environmental factors such as water, salt, alkali, ultraviolet light, elevated temperatures, and freeze-thaw cycles could potentially degrade the components of FRP composites and the interfaces in both the external and internal applications of FRP. Alkaline exposure is identified as the most damaging condition, with CFRP displaying superior performance in aggressive environments, while GFRP exhibits resilience across various conditions. The significance of the glass transition temperature (Tg) is crucial for both applications (i.e., internal and external); FRP composites could lose their bond with concrete when subjected to service temperatures approaching their Tg. Guidelines highlight the importance that FRP composites should not be utilized in environments where the service temperature exceeds 27°F below their Tg.

### 4.4.4 Fabrication and Workmanship

The sources of deficiencies associated with the fabrication and workmanship errors in the FRP composites can be classified as manufacturing errors (anywhere off-site), termed as installation/casting errors (on-site), or during transportation, storage, and handling (Table 4.6).

**TABLE 4.6**
**Fabrication and workmanship factors introducing defects in FRP-RSC elements [266]**

| Manufacturing errors | Transportation, storage, and handling errors | Installation errors |
| --- | --- | --- |
|  |  |  |
| FRP composites are manufactured by a variety of methods, and the defects can be introduced at any stage of manufacturing anywhere within the matrix, fiber, and interface of the composite. They are more susceptible to variations in material and geometric properties due to their complicated manufacturing processes, as can be seen in the figure above, which shows an example of incorrectly bent bars by the manufacturer. | FRP composites may bend, break, and get coated with dirt, oil, or other materials during transportation, as shown in figure above (GFRP bars broken due to coiling on a reel with an unacceptably small diameter). Similarly, improper lifting, dropping, and dragging during handling introduce deficiencies in them. Care should be taken for their storage as prolonged exposure to UV rays and temperatures above 120°F deteriorates their mechanical properties. | Workmanship problems, faulty equipment used for their application, unfavorable temperature or moisture conditions, inconsistent surface preparation, improper mixing or quantity of resins, uneven distribution of resins, inadequate curing of resins, inconsistent layering and sagging of FRP plies (as shown in figure above), and many other deviations from the suggested method of their installation could all result in errors during the installation of external FRP. |

## 4.4.5 Etiology of Deficiencies in FRP-RSC Elements

The etiology of deficiencies (cause and effect) shown in Table 4.7 establishes a rational relationship between the observed or expected defects in FRP-RSC elements and their causes. This table can be used to trace back the cause of a defect observed during the inspection of the FRP-RSC elements. Further, deficiency etiology dedicated to FRP-RSC elements aids in effective and accurate detection of deficiencies and analyzing their impact on serviceability and structural performance.

**TABLE 4.7**
**Etiology of deficiencies**

| Deficiencies \ Source of Deficiencies | Design Factors | Mechanical Factors | Environmental Factors | Fabrication Errors | Workmanship |
|---|---|---|---|---|---|
| A.1 Blisters | | | X | X | X |
| A.2 Wrinkling | | | | X | X |
| A.3 Scratches | | X | | | X |
| A.4 Discoloration | | X | X | | |
| A.5 Fiber exposure | | X | X | | X |
| A.6 Voids at fiber–matrix interface | | | X | X | X |
| A.7 Debonding at fiber–matrix interface | X | | X | X | X |
| A.8 Delamination between composite layers | X | X | X | X | X |
| A.9 Cracks in FRP | X | X | | | X |
| A.10 Impact damage | | X | | | |
| B.1/D.1 Debonding between FRP and concrete | | X | X | | X |
| B.2/C.1/D.2 Voids between FRP and concrete | | X | X | | X |
| E.1 Cracks in concrete substrate | X | X | | | |
| E.2 Voids in concrete | | | X | | X |
| E.3 Concrete delamination | X | X | | | |
| F.1 Cover separation | X | X | | | |
| F.2 Corrosion in steel reinforcement | | | X | | |
| G.1 Loss of cross-sectional property | | | X | | |
| H.1 Debonding of internal FRP reinforcements | | X | X | | |
| H.2 Slippage | | X | X | | |
| I.1 Cracks in FRP reinforced concrete | X | X | | | |

# 5 NDT Methods for Detecting Defects in FRP-Reinforced/Strengthened Concrete

Nondestructive testing, in general, is defined as "an examination, test, or evaluation performed on any type of test object without changing or altering that object in any way, in order to determine the absence or presence of conditions or discontinuities that may have an effect on the usefulness or serviceability of that object" [272, 273]. NDT methods have been increasingly used for quality control, quality assurance, and quality assessment of both new and old structures [274, 275].

This chapter will review the nondestructive testing techniques that could potentially be applicable to FRP inspection through a literature survey of past studies, applications, and research projects. For each NDT technique, its theoretical background and principle will be discussed first. Subsequently, the past applications of the individual NDT methods for FRP inspection will be listed, focusing on the conclusions made from each of those previous experiences. Finally, the advantages and disadvantages of each of these methods will be addressed for assessing their applicability to FRP inspection. All these findings will serve to evaluate the suitability of each method for FRP inspection.

There have been some noticeable studies on the application of visual testing, vision-based techniques, tap testing, impact echo testing, microwave testing, ground-penetrating radar, ultrasonic testing, infrared thermography, acoustic emission, laser testing, and radiographic testing for nondestructive evaluation of externally bonded FRPs. Similarly, other NDT methods such as modal testing, load testing, and robotic inspection could have great potential to be used for FRP inspections, owing to providing increased efficiency. However, the search for the NDT methods that can be used for embedded FRP rebars did not return many results, confirming the fact that there has been little research performed in this area and that the task of detection is significantly challenging.

## 5.1 THE NEED FOR NDT IN FRP INSPECTION

Defects and damages that could be expected for FRP-reinforced/strengthened concrete (FRP-RSC) elements were discussed in the previous chapter. These include defects/damage within FRP, within concrete, and at the interface of FRP and


DOI: 10.1201/9781003646969-5

concrete. Most notable damage for the performance of FRP-RSC elements, however, has proven to be the issues at the concrete–FRP interface, as described in more detail in the following.

The bond between the embedded FRP rebars and the concrete is crucial for the optimum performance of FRP-reinforced concrete structures. This interfacial bonding consists of contributions from (a) chemical adhesion, (b) mechanical interlock, and (c) frictional resistance where the first mechanism accounts for the interfacial bond/adhesion and the latter two account for the pullout strength [276]. The deterioration of the interfacial bonding due to environmental or load factors may result in debonding between the FRP rebars and the concrete, which contributes to the structural capacity of FRP-reinforced civil structures.

Similarly, in the case of externally bonded FRP composites, the additional capacity to resist the flexural and shear stresses depends on the composite action of the whole system, which in turn relates to the quality/integrity of the bond between the FRP composites and the concrete substrate. Hence, it is crucial that any debonding, delamination, or voids between the externally applied FRP composites and the concrete surface be avoided as much as possible. However, the bonding may not meet the desired quality due to various factors during the construction/installation process or service life, which again compromises the satisfactory performance of the structures strengthened with FRP.

Regardless of the type, damage or defects negatively affect the structural integrity and serviceability of FRPSRC elements and, therefore, it is essential to have a set of NDT tools capable of detecting these defects for both the embedded FRP rebar and the externally bonded FRP composite applications. With the increase in the use of innovative materials such as FRP, the need for new or study on the feasibility of existing NDT methods becomes all the more apparent to ensure construction quality and continued structural integrity. The need for NDT is even more urgent because of the following two factors [277]. First, the failure of structures reinforced with FRP rebars is not as ductile as conventional constructions and hence it is vital to detect signs of potential failure as early as possible. Second, concrete structures strengthened with externally bonded FRP composites are covered and cannot be inspected as readily. Thus, this requires NDT methods that can inspect beyond the FRP wraps or laminates to examine the internal integrity of the element.

## 5.2 NDT METHODS FOR FRP INSPECTION

### 5.2.1 Visual Inspection (VT)

Visual inspection is the most common, versatile, and simplistic NDT method [25, 278, 279] that can be used for inspecting FRP-reinforced/strengthened concrete elements. There are two ways in which a visual inspection can be conducted: (1) direct visual testing and (2) remote visual testing (see Figure 5.1) [265]. In direct visual testing, the inspection surface is within direct view of the human eye, although tools such as cameras, mirrors, binoculars, magnifying lenses, and flashlights may be used to improve visibility. Additionally, devices like measuring tapes, calipers, gauges,

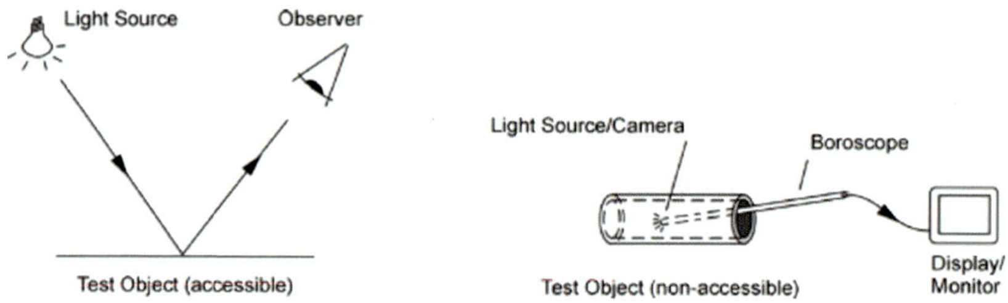


**FIGURE 5.1** Direct and remote visual testing [265]

and micrometers are employed to record quantitative data on detected defects. In contrast, remote visual testing uses equipment such as borescopes, fiberscopes, and conventional video cameras to obtain clear, interpretable images of surfaces that cannot be seen directly. Regardless of whether the method is direct or remote, the accuracy of visual inspection results largely depends on the inspector's visual perception and experience.

Visual inspection is an inexpensive, fast, and effective method that can be used for both global and local inspection of structures/elements. It can be used to check for the obvious and outstanding anomalies on a global basis by taking the appearance of the sound regions as a baseline for comparison. After inspecting for defects/flaws globally (over large areas), a detailed distinction of the local flaws can also be carried out in no time. It usually does not require dedicated equipment, and data is available in real time for instantaneous interpretation. Defects such as cracks, delamination, fractures, blisters, debonding, moisture accumulation, fiber misalignment, discoloration, or any other surface defects may be detected by visual inspections. However, visual inspection cannot detect subsurface defects located deep within the elements as reliably as it detects surface defects. Experienced inspectors may be able to diagnose internal subsurface defects simply from the surface observations by interpreting them as symptoms of internal anomalies and may even recommend further subsurface inspections using other more sophisticated NDT methods.

From the visual abilities of the human eye to the judgment calls for the interpretation of data, there is a human element involved in the process of visual inspection. Hence, one of the major limitations of this method is that it is subjective, and observations may vary depending on the visual perceptions of individuals. Similarly, being limited to detecting only surface defects is another disadvantage of using visual inspection as a method for NDT.

In summary, visual inspection is fast, simple, and cost-effective, and based on the findings of visual inspection, decisions can be made on the necessity of further inspection with more detailed NDT methods. Visual inspection can be considered the most fundamental and primary NDT tool as it sets the baseline for other NDT methods and should be performed before any further NDT as a supplementary method. However, it is also important to be wary of its limitations; it is unreliable to give the big picture of the defects in full depth but rather only detects defects on the surface.

### 5.2.2 Tap Testing (TT)

The tap testing method is based on the principle that the local stiffness of a material in a defective area differs from that in intact regions [280]. In defective zones, the reduced stiffness alters the excitation frequency when the surface is struck [281]. In practice, this technique involves tapping the test object with a coin or hammer, as illustrated in Figure 5.2, and listening for changes in the sound's frequency. A sharp, high-pitched sound indicates that the material is in good condition, with strong bonding either at the FRP–substrate interface or between the concrete's interlaminar layers. Conversely, a dull, hollow sound suggests the presence of defects such as voids, debonding, or delamination [28, 279, 282].

Tap testing is a widely used technique for identifying subsurface defects such as debonding, delamination, and voids between FRP wraps and the underlying concrete [283–285]. This fast, low-cost NDT method is particularly effective for inspecting FRP-strengthened structures [279, 286] and is most commonly applied to detect delamination [287, 288]. While visual inspection is limited to surface flaws, tap testing can detect medium to large subsurface defects. However, due to its limited penetration depth, only a few millimeters (5–10 mm), primarily it is still considered a

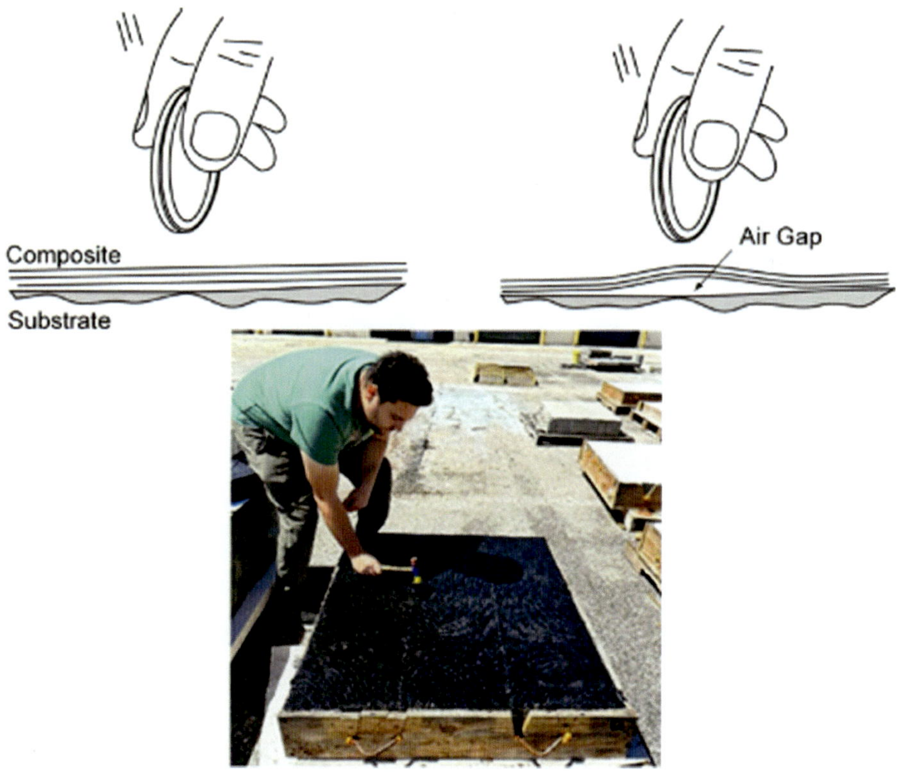


FIGURE 5.2 Coin tap testing [265] and hammer tap testing

surface inspection method [284]. It offers real-time results for detecting discontinuities such as delamination, debonding, and voids in thin composites, and it can cover large areas in a short time [287]. Moreover, it requires minimal operator training, further adding to its practicality [25].

The reliability of the tap testing method is limited because it relies heavily on the inspector's hearing ability and experience in interpreting signals, making the results inherently subjective—a major drawback of the technique [280]. Variations in tapping force, angle, and the type of device used can further affect consistency [282]. Additionally, frequency changes caused by geometric variations in the structure [25] or by ambient noise [279] may be mistakenly identified as defects. The method also cannot provide quantitative or objective data on the size or depth of defects, and fine flaws such as porosity or fiber breakage often go undetected due to the limited sensitivity of human hearing [287, 288].

The limitations of conventional tap testing have been addressed with the development of commercial digital tap testing (DTT) devices, as illustrated in Figure 5.3. Halabe, Joshi, and Gangarao [284] demonstrated that DTT can detect debonding as small as 1.5 × 0.75 in. (1.1 in²), although ACI 440.2R-17 [5] states that debonds smaller than 2 in² generally do not require special attention if the affected area is within 5% of the total bond area. In DTT, an impact hammer is used to tap the test surface, and the contact time (impact duration) is recorded. Defective areas produce a broader force–time pulse or longer contact time, whereas sound areas yield shorter readings. By providing quantitative results, DTT eliminates the subjectivity inherent in manual tap testing and enables more consistent and reliable defect detection.

**FIGURE 5.3** Digital tap testing devices

### 5.2.3 Impact Echo Testing (IE)

Impact echo testing works by generating stress or mechanical waves, typically using a hammer or steel sphere, that travel through the structural member and reflect back when they encounter changes in acoustic impedance caused by discontinuities. A transducer placed near the impact point records the reflected waves as a time-domain signal (voltage versus time), which is often converted into a frequency-domain signal (amplitude versus frequency spectrum) to simplify analysis [279, 289–291]. By examining these signals, the extent and location of subsurface defects can be determined.

The impact echo testing method enables rapid inspection for locating defects, which can then be examined in greater detail using other comprehensive NDT techniques. It is applicable for both quality control and defect detection in FRP composite installations [292]. Maerz, Galecki, and Nanni [293, 294] conducted impact echo testing using an Olson Instruments impact echo tester, modified with an air-coupled receiver, to identify delamination in externally bonded FRP laminates on a bridge abutment (Figure 5.4). The inspection was performed on a 1 × 1 in. grid, and the results showed that the method was highly effective in detecting delaminations in FRP applications.

Unlike ultrasonic testing, which relies on high-frequency stress or sound waves and is often affected by significant signal attenuation and noise, impact echo testing uses a hammer or steel ball to generate high-energy, low-frequency stress waves (typically 2–20 kHz) capable of penetrating deep into the material [285, 291, 295]. This impact-based approach also eliminates the need for a bulky transmitting transducer, which is necessary in ultrasonic testing [289, 296].The method is well-suited for detecting cracks, voids, and delaminations in plate-like structures, where such defects are usually parallel to the surface [291], making it highly applicable to external FRP systems. Since inspections can be carried out from a single accessible side, it removes the challenge of gaining access to both sides of the structure [287]. Additionally, it offers notable advantages in accuracy, repeatability, and speed for crack detection [297, 298].

**FIGURE 5.4** Impact echo testing [293]

However, impact echo testing also has certain limitations. Because it employs low-frequency stress waves, it has difficulty detecting small cracks and fine discontinuities, which poses a significant limitation in terms of resolution [291]. In addition, current IE equipment can only evaluate members up to 40 inches (1 m) in thickness, and the method requires skilled, experienced operators to ensure accurate results—factors that further constrain its application [289].

### 5.2.4 Microwave Testing (MW)

Microwaves are electromagnetic waves with frequencies between 0.3 and 300 GHz, corresponding to wavelengths ($\lambda = c/f$) ranging from 40 inches (1 m) to 0.04 inches (1 mm) [299]. Microwave testing works by analyzing the phase, frequency, and amplitude differences between transmitted and received microwave signals as they pass through a test medium [300]. When these signals interact with a layered material such as FRP composites, they are either transmitted or reflected depending on the differences in dielectric properties at the interfaces between layers [301]. By examining the electrical characteristics of the reflected signals, it is possible to infer material properties at these boundaries and detect subsurface defects such as delamination, debonding, and voids [278].

The microwave testing technique has been applied to detect defects in FRP composites, as well as at the FRP–concrete interface [302–305]. It is capable of determining the depth, location, geometry, and dimensions of flaws in FRP systems [279, 291]. For example, J. Li and Liu [306] used microwaves to identify air voids between FRP layers and concrete substrates. Likewise, Stephen et al. [307] successfully detected delaminations smaller than 20 mm × 20 mm in CFRP laminates bonded to both the abutment and the girder of a bridge.

Conductive materials block microwaves, making microwave testing most suitable for non-conductive or dielectric materials [282, 300]. Its application in CFRP composites is limited because these materials behave like conductors, although it can effectively penetrate thick dielectric materials such as GFRP [282, 300]. While microwave testing offers the benefits of non-contact and one-sided inspection, its penetration depth is restricted by the dielectric properties of the FRP composite. Consequently, radar-based methods such as ground-penetrating radar (GPR), which provide greater penetration depth [291], are often preferred over microwave testing [282]. Due to these limitations, GPR has seen wider use than microwave testing for nondestructive evaluation, particularly in CFRP-rehabilitated structures [265]. The GPR technique is discussed in detail in the next section.

### 5.2.5 Ground-Penetrating Radar (GPR)

Ground-Penetrating Radar (GPR) is a real-time NDT technique used to evaluate internal damage in structural elements. It works on the principle that when electromagnetic waves encounter an interface between two materials with different dielectric constants, part of the wave is reflected back (Figure 5.5) [289, 308]. In practice, GPR transmits electromagnetic (radio) waves into the material and records the waves

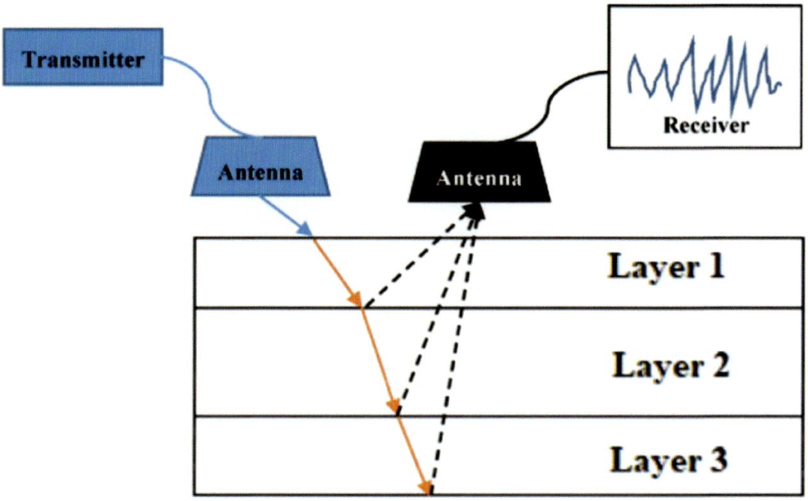


**FIGURE 5.5** Reflection of incident waves at different interfaces [311]

reflected from discontinuities within it. These discontinuities can be material interfaces, such as the concrete–rebar interface, or subsurface defects like voids, cracks, debonds, and delaminations (e.g., concrete–air or concrete–water interfaces), all of which exhibit distinct dielectric properties. The nature of these discontinuities can be inferred from the amplitude (intensity) and velocity of the reflected waves, both of which depend on the change in dielectric properties at the interface [309, 310].

The penetration depth and resolution of GPR depend on the frequency of the transmitted pulse. Lower frequencies provide greater penetration but lower resolution, while higher frequencies offer improved resolution for detecting small defects, though only within shallower depths [289, 312]. Technological advancements and the wider availability of commercial GPR antennas in various frequency ranges have expanded its use for inspecting subsurface conditions in elements with thin FRP composite layers [279, 291].

A feasibility study by Jackson, Islam, and Alampalli [313] on FRP-wrapped bridge columns demonstrated that GPR can detect defects or deterioration at nearly any depth, provided the appropriate antenna frequency is selected. It found that shallow defects, such as debonding and delamination, can be effectively identified using a high-frequency antenna of around 2 GHz, whereas a lower-frequency antenna of about 1 GHz was less effective for such defects. Similarly, a 1.5 GHz ground-coupled antenna performed well in detecting water-filled defects as small as $50.8 \times 50.8 \times 1.5$ $mm^3$ but was less accurate for identifying air-filled defects in FRP-wrapped concrete cylinders [308] and GFRP bridge decks [314].

Furthermore, Yazdani et al. [315] evaluated four construction-related factors, surface voids, surface wetness, surface cleanliness, and the orientation of CFRP application (upward vs. downward) that can affect bond quality during CFRP installation. Of these, GPR was only effective in reliably detecting surface voids.

One key advantage of GPR (radar/electromagnetic wave methods) over stress wave techniques such as ultrasonic testing and impact echo is its ability to penetrate beyond concrete–air interfaces, enabling the inspection of features located beneath these boundaries [289]. In contrast, stress wave methods are more sensitive to detecting concrete–air interfaces, as the waves are almost entirely reflected when encountering air. GPR also offers greater penetration depth compared to other NDT methods like infrared thermography. Additionally, the use of air-coupled GPR, which does not require direct contact with the inspection surface, speeds up the inspection process by eliminating the need for coupling procedures.

There are several limitations to using GPR for FRP-reinforced or strengthened concrete elements that should be taken into account. One major drawback is its relatively high cost compared to other NDT methods, such as infrared thermography [315]. As noted earlier, GPR is not effective in detecting air-filled defects and should therefore be used in conjunction with other techniques—such as infrared thermography [314] or stress wave methods—to identify such flaws. Furthermore, due to the limited amount of research on GPR applications for FRP-reinforced or rehabilitated structures, it has yet to be firmly established as a consistently reliable NDT method. Nonetheless, based on the reviewed literature, GPR shows promising potential for specific applications, particularly in detecting debonds and water-filled delaminations in FRP–RSC elements.

### 5.2.6 Ultrasonic Testing (UT)

Ultrasonic waves are stress or sound waves with frequencies above the human audible limit of 20 kHz. Ultrasonic testing operates on the principle that when these waves encounter an interface between two materials with different acoustic impedances, part of the wave is reflected. Defects, typically containing air or water, have acoustic impedances that differ significantly from those of the surrounding intact material [265]. This contrast in acoustic properties at the interface between defective and sound regions is the key factor enabling ultrasonic detection of discontinuities. The method has proven highly effective for identifying defects in both concrete and composites, as the coefficient of reflection, a measure of the proportion of acoustic energy reflected at an interface, is close to 1 for both concrete–air interfaces [289] and composite–air interfaces [265]. This indicates that defects at such interfaces cause almost total wave reflection.

Ultrasonic testing can be performed by either measuring waves that pass completely through the material or by analyzing waves reflected from internal defects. The first approach, known as the ultrasonic through or direct transmission method, also referred to as the ultrasonic pulse velocity (UPV) method (Figure 5.6), involves measuring the travel time of ultrasonic compression waves over a known path length from one side of the member to the other. The wave velocity, calculated by dividing the member thickness by the measured travel time, is then used to assess the quality and uniformity of the concrete, indicating the possible presence of defects.

However, the UPV method has notable limitations: it requires access to both sides of the member and can only confirm the existence of defects, without providing

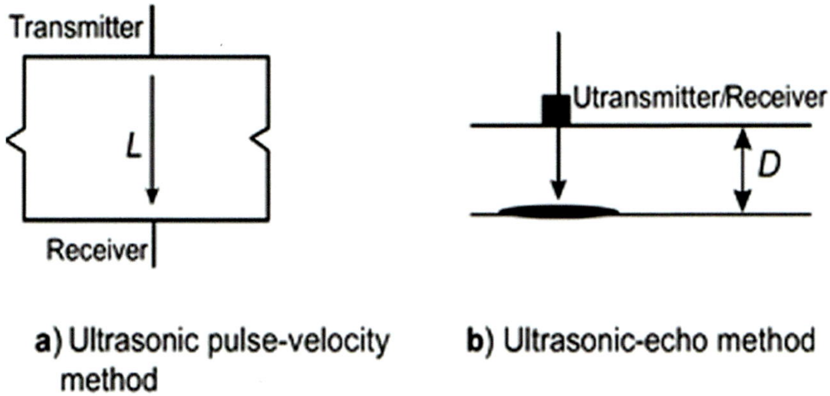


**FIGURE 5.6** Ultrasonic pulse velocity method and ultrasonic echo method

information on their depth. In contrast, the ultrasonic echo method (Figure 5.6) overcomes these constraints by using a transducer to generate ultrasonic waves that propagate through the material and are reflected back by discontinuities such as flaws or interfaces.

Numerous studies have explored the application of ultrasonic testing for inspecting various FRP–concrete interfaces. Kundu et al. [316] demonstrated that both longitudinal (P) waves and Lamb waves can be used to detect delaminations at the interface between concrete and GFRP plates. Bastianini, Di Tommaso, and Pascale [317] employed the ultrasonic echo method to assess bonding defects in external FRP strengthening systems by analyzing the amplitude of reflected waves. La Malfa Ribolla et al. [318] introduced an ultrasonic technique for evaluating FRP–substrate bond quality using Equivalent Time Length (ETL), a novel damage-sensitive parameter that eliminates the need for direct coupling between transducers and the inspected material. Concu and Trulli [319] studied and investigated the use of the UPV method to assess adhesion quality between FRP and its substrate, concluding that both direct and semi-direct transmission modes are effective for detecting adhesion defects.

Ultrasonic testing is a fast and field-friendly technique with high resolution for flaw detection [320]. It is useful for measuring material thickness and identifying cracks, voids, and delaminations [292]. In FRP applications, it can also detect a range of discontinuities, including debonds, resin inconsistencies, broken fibers, impact damage, moisture ingress, and other subsurface defects [25]. While ultrasonic testing can effectively penetrate materials and locate flaws both within composites and at the FRP–concrete interface, a key drawback is its reliance on highly trained operators for both execution and data interpretation [25, 291]. The UPV method, in particular, provides immediate, detailed, and accurate results with minimal preparation, but it requires adequate access, skilled personnel, and is best suited for homogeneous materials [279].

### 5.2.7 Phased Array Ultrasonic Testing (PAU)

Ultrasonic waves are generated using devices called ultrasonic transducers, which convert electrical or optical signals into ultrasonic waves and vice versa (Figure 5.7). The earlier discussion on ultrasonic testing focused on single-element transducers; however, when two or more individual transducers are arranged to function together, they form an array. If this array is operated so that its elements are pulsed separately, with slight time delays (phasing) either electronically or physically, the resulting constructive and destructive interference produces an angled longitudinal wave. This configuration is known as a phased array [321].

Unlike conventional ultrasonic testing, phased array ultrasonic (PAU) testing enables signal focusing at specific locations and angles, offering a significant advantage for inspecting composite materials with anisotropic structures, which often pose challenges for signal interpretation [322, 323]. Boychuk, Generalov, and Stepanov [324] concluded that PAU can detect all common FRP defects while accurately determining their size and location. Similarly, Meola et al. [325] demonstrated its effectiveness in identifying defects and impact damage in CFRP composites, successfully pinpointing both size and position. Taheri and Hassen [322] further showed that PAU outperforms conventional ultrasonic testing, with the ability to detect flaws as small as 0.8 mm and achieve penetration depths of up to 25 mm.

The advantages of PAU include a scanning rate five to ten times faster than conventional UT, improved imaging through the use of multiple angles and frequencies (reducing the need for extensive interpretation), higher resolution, beam-focusing capability, and enhanced reliability, portability, and mobility [326, 327]. However, as a relatively new technique compared to traditional NDT methods, portable PAU units still face limitations, including uncertainties in application since the method has yet to be fully validated [327].

**FIGURE 5.7** Phased array ultrasonic device

### 5.2.8 Infrared Thermography Testing (IR)

Infrared thermography works on the principle that anomalies within concrete have different thermal properties, specifically thermal conductivity, compared to the surrounding intact material (Figure 5.8). When heat flows into or out of the test object, the variation in heat transfer rates between defective and sound areas creates localized temperature gradients. These gradients appear as surface temperature differences, which can be measured to identify subsurface anomalies.

Infrared thermography (IRT) has been used to detect defects in externally bonded FRP laminates and laminates used for strengthening concrete members [289]. Jackson, Islam, and Alampalli [313] found that IRT can effectively identify debonding, blisters, and shallow defects such as delamination in concrete columns with externally applied FRP. Galietti et al. [328] explored its use for detecting defects at the concrete–FRP interface and concluded that it is one of the cheapest and quickest techniques available for this purpose. Additionally, research on the quantitative assessment of defects in FRP-bonded concrete has shown that IRT can not only locate defects but also estimate their size and approximate depth [329, 330].

The advantages and limitations of infrared thermography (IRT) are key considerations in evaluating its suitability as an NDT method for inspecting FRP-strengthened concrete. IRT is a non-contact technique capable of detecting subsurface defects over large surface areas in a short time, offering broader coverage than many other NDT methods [289]. It can also identify localized defects while being more cost-effective than several alternative techniques [265]. Yazdani, Beneberu, and Riad [315] found that IRT could clearly detect all four construction-related factors, surface voids, surface wetness, surface cleanliness, and the orientation of CFRP application, that affect bond quality during installation. Additionally, IRT enables real-time inspections and data interpretation, allowing for immediate structural assessment.

**FIGURE 5.8** Infrared thermography [288]

However, the method has limitations. It is less effective at detecting water-filled defects [314] and defects located deep within the concrete [313]. Another drawback is its reliance on environmental conditions to establish the necessary heat flow, which produces the surface temperature variations required to reveal subsurface anomalies.

### 5.2.9 Acoustic Emission Testing (AE)

In acoustic emission (AE) testing, elastic waves generated by sudden stress redistributions, caused by phenomena such as moving dislocations, crack initiation, growth, and propagation, fiber breakage, debonding, and plastic deformation, are detected using AE sensors (Figure 5.9). These elastic waves, known as acoustic emissions (AEs), originate within the material itself rather than being introduced from an external source. Detecting these emissions provides information about the location and nature of discontinuities within the material. However, AE testing is not considered a true NDT method because it requires the application of loads that induce cracks and strains, resulting in permanent, non-reversible damage to the structure.

Mirmiran, Shahawy, and El Echary [331] evaluated the feasibility of using the acoustic emission (AE) technique as an NDT tool to inspect hybrid columns, CFRP tubes filled with concrete, and established correlations between AE parameters and the stress states in the concrete. In another study, Gostautas et al. [332] applied AE to assess the structural performance of multiple GFRP bridge deck panels with depths ranging from 6 to 30 inches. Carpinteri, Lacidogna, and Paggi [333] used AE to monitor and analyze crack progression leading to FRP debonding in a reinforced concrete beam retrofitted with FRP laminates. Similarly, Degala et al. [334] performed AE tests on concrete slabs strengthened with CFRP strips and found that the method could identify different sources of damage, particularly distinguishing between concrete cracking and CFRP debonding, which were the two primary failure

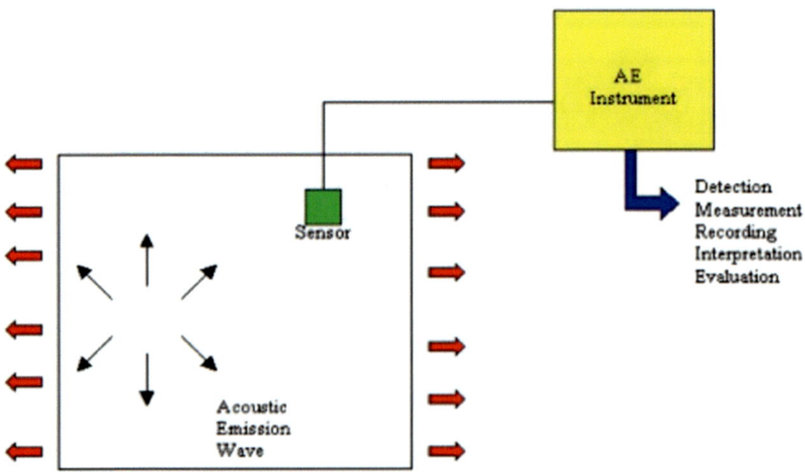


**FIGURE 5.9** Acoustic emission testing [25]

modes observed in their laboratory specimens. Overall, these studies have reported promising outcomes for the use of AE in monitoring various FRP applications.

Acoustic emission (AE) testing enables rapid, large-scale inspections by employing multiple AE sensors that can be permanently mounted with process control, eliminating the need for specimen cleaning or disassembly [279, 320]. Once installed, the method is less labor-intensive than many other NDT techniques and can be conducted without interrupting the service of the structure being monitored [291]. AE also excels at detecting a wide range of defects associated with fatigue loading, including fatigue cracks, fiber fractures, matrix micro-cracks, fiber–matrix debonding, and delamination [320].

Unlike most other NDT methods that rely on externally introduced stress waves, AE detects stress waves generated within the material itself [321], allowing precise identification of defect origins. Furthermore, while conventional NDTs typically detect existing geometric discontinuities, AE can monitor dynamic processes such as movement and strain, making it possible to track defect growth rates over time [25].

While AE testing offers several advantages, its application is limited by certain challenges. One major drawback is the difficulty in distinguishing stress waves generated by a specific damage source from those produced by unrelated sources, such as vehicular traffic [292]. This makes it essential for operators to have a high level of expertise in correlating AE data with the corresponding damage mechanisms [320]. Additionally, noisy environments can introduce false signals, making effective filtering and noise reduction crucial during data collection [279].

Another limitation is that defects may go undetected if the applied loading is insufficient to trigger a significant AE event [286]. Moreover, although AE can detect defects, locate their origin, and estimate their approximate position, it cannot determine their exact type or size. Therefore, complementary NDT methods, such as ultrasonic testing, are often needed to provide complete quantitative information about the defects [291].

### 5.2.10 Laser Testing Method (LT)

Laser-based nondestructive testing methods can be performed using either interferometry or vibrometry, both of which measure time-varying displacements and particle velocities with an optical beam. Interferometry can be divided into two main types: holography and shearography. In holographic interferometry, defects or deformations in an object appear as anomalies in the fringe pattern obtained by recording and comparing holograms of the object in two states, before and after load application, or after subjecting it to a small mechanical, thermal, pneumatic, or vibrational excitation [335]. In contrast, shearography detects defects by analyzing fringe contours, which represent displacement derivatives, from images of the object in its deformed and undeformed states [335].

Laser vibrometry, in contrast, measures the velocity or displacement of an object's surface vibrations using the Doppler effect. The acoustic–laser vibrometry method is a relatively new technique for detecting delamination and voids in FRP composites.

In this approach, the test object is excited acoustically, and a laser vibrometer records its vibration signatures or surface velocities [336, 337]. By analyzing these vibration signals, the presence as well as the approximate size and shape of defects can be identified, since defective areas in FRP composites exhibit different vibration patterns compared to the surrounding intact regions (Figure 5.10).

Taillade et al. [338, 339] demonstrated the feasibility of using shearography to evaluate bond defects between externally applied CFRP and concrete substrates through a case study on a bridge over the Doubs River in France. Yang [340] applied digital shearography to detect defects and delaminations in a CFRP specimen under tensile stress, a GFRP plate under thermal stress, and a GFRP honeycomb panel under partial vacuum stress. Qiu and Lau [341, 342] experimentally confirmed the viability of the acoustic–laser technique for detecting defects in FRP-bonded concrete and concluded that it can also quantify defect size. J. G. Chen, Haupt, and Büyüköztürk [343] reported that acoustic–laser vibrometry could detect a shallow crack in concrete beneath externally applied FRP. Additionally, Büyüköztürk [343] experimentally validated the use of the acoustic–laser technique for remotely detecting debonding in GFRP-confined concrete specimens.

Laser-based nondestructive testing methods offer key advantages such as noncontact operation, high sensitivity, and a high detection rate [344]. Unlike conventional contact accelerometers, they do not add mass to the test object, ensuring that measurements are not influenced by the added weight of the testing device or procedure [336]. The acoustic–laser technique, in particular, provides benefits including remote inspection capability, the ability to detect small structural elements, low

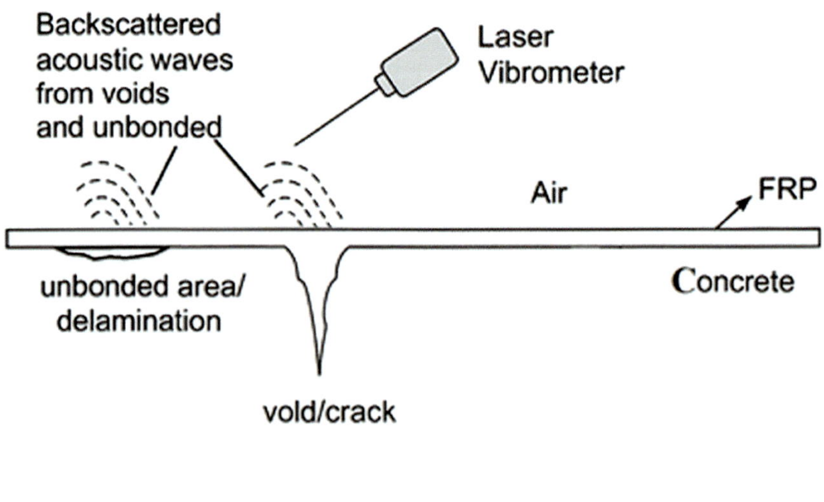


**FIGURE 5.10** Acoustic-laser vibrometry

**TABLE 5.1**
**Advantages and disadvantages of laser-based NDTs [345]**

| Technique | Advantage | Disadvantage |
|---|---|---|
| Laser interferometry | • Non-contact method<br>• High sensitivity<br>• High detection rate | • Require laborious analysis of the results<br>• Expensive device |
| Acoustic-laser vibrometry | • Non-contact method<br>• Less harmful to people<br>• Capable of detecting small defects | • Hearing impairment to people<br>• Eyes protection is needed<br>• Noise may affect the effective signals |

maintenance and operating costs, and overall light weight and mobility [342]. A summary of the advantages and limitations of laser-based nondestructive testing methods, as presented by Lau and Qiu [345], is provided in Table 5.1.

### 5.2.11 Radiographic Testing (RT)

Radiography is the sole NDT method that works on a variety of materials, and since it has been in use for over a century, it can be considered one of the most mature methods as well [321]. The principle behind radiographic testing is recording the intensity of high-energy electromagnetic radiation, such as X-rays and gamma rays, on a photographic film after passing them through the test material. As the radiation waves pass through the specimen, different materials will have different wave attenuation rates (or different absorption rates); for instance, steel attenuates more of the X-rays and gamma rays than concrete, and concrete more than air, as denser material blocks more of the radiation [289, 296]. Hence, the differences in the intensity of the radiation that can be seen as shadows or bright spots [12] in the photographic film give information about the internal features of the member that is being tested. Light areas in the film indicate high-density material, such as reinforcements, and the dark areas refer to the low-density features, such as voids or delamination [289].

Radiographic tests can clearly detect defects like delaminations in the FRP, as there would be variations in the absorption of the radiations between the sound and defective regions [346]. However, if the orientation of the delamination is perpendicular to the radiations, these defects won't be detected in a 2D scan [320]. Defects that cause significant density variations such as excess density of fiber or matrix, resin variations, impact damage, delamination, debonding, foreign contamination, cracks, and voids can be easily detected by radiography [12]. It is also useful in detecting non-uniform fiber distribution, broken fiber, poor fiber weaving, and misorientation in fibers such as fiber wrinkles. It can be used for both quality control of composites before service and for inspections for defects during service [12].

Radiography can be used for exact detection of defects owing to its high resolution and sensitivity; however, the resolution depends on the orientation of the defects

[25, 291]. By using the source and the detector/receiver at different positions, accurate 3D images of the internal structures (tomographic image) of the test member can also be generated [347]. However, the major disadvantage of radiographic techniques is the health risk associated with it. Improper handling and storage of radiographic instruments can invite serious health hazards [25] as the human tissues will be seriously damaged under exposure to the radioactive waves used in this method [12]. Hence, due to this fact, radiographic testing should only be conducted by qualified personnel with a high level of skill and adequate knowledge of safety precautions and data evaluation [25]. The radiographic technique already has a high equipment cost [291], and the requirement of safety precautions further adds to the cost of its operation [12]. Moreover, this method has another drawback of requiring access to both sides of the test object for placing the source and receiver on opposite sides.

## 5.3 GLOBAL STRUCTURAL RESPONSE TESTING

Global structural response testing evaluates the overall performance of a bridge/structure as a system rather than as individual components. Nevertheless, changes in global behavior (static and dynamic) can be due to the presence and location of damage.

### 5.3.1 Modal Testing

The basic principle behind modal testing or modal analysis is to evaluate the condition of a structure by monitoring the changes in its dynamic response/behavior (typically, modal frequencies and modal shapes). The vibration characteristics or the dynamic response of a structure change with degradation as it alters the material properties, mass, and stiffness of the structure [282]. Thus, the dynamic response of an FRP-RSC bridge after deterioration will be different from that of its original state, and by comparing the difference, information about damage in the structure can be determined [346]. For instance, decrease in modal frequency of the existing structure from that of the original structure will denote a decrease in the stiffness of the structure [265]. In order to perform this test, an array of accelerometers is installed at multiple locations in the structure to obtain sufficient data on the required modal parameters.

Vibration-based modal analysis is one of the most widely used techniques in health monitoring and damage detection of structures based on their global response [298]. The use of modal analysis has been prominently increasing for monitoring long-term deterioration in civil structures [265]. This method has been successfully used by Caltrans for monitoring and non-destructive evaluation of bridges reinforced and strengthened with FRP composites [25]. Similarly, Alampalli [348] used modal analysis to study the dynamic behavior of the first FRP composite bridge built in New York state, and Guan, Karbhari, and Sikorsky [349] used it to determine the long-term performance of an FRP highway bridge. Recently, modal analysis has been mostly used for investigating the efficiency of CFRP rehabilitation applications [265].

Modal testing, however, has some drawbacks. The changes due to discontinuities within the FRP composites may not be significant enough to contribute to the overall modal response of the large structures. This shows the limitation of the method in terms of sensitivity for use in large structures. Moreover, modal analysis cannot provide real-time and instantaneous assessment of the structural integrity as it requires complex data interpretation off-site. Similarly, this method requires the modal response of the original state of the structure before damage occurs. Although modal analysis is very effective in conducting global structural response analysis, it is not as effective in the detection of localized damage in externally bonded FRP composites [265].

### 5.3.2 Load Testing and Response Measurement

The principle of detecting defects using load testing is similar to that of modal testing, in which a reduction in stiffness reflects the presence of deterioration [282]. It is based on the assumption that the stiffness of the structure reduces with the degradation of the structure. Thus, a change in static response, such as an increase in deflection of a structure under externally applied load, would indicate that there might have been some degree of material degradation. Such variations in static response are sensed by the load testing system that measures strain or deflection using strain gauges and other deflection-measuring devices/sensors installed at several strategic points along the structure. In order to obtain readings that are sufficient to be reflected in the overall response of the structure for accurate and reliable damage detection, a large number of different loading configurations are required, and displacement or strain measurements should be taken under several known static load cases. The live load used for load testing should be applied at predefined patterns on a bridge, and its gain in performance (with respect to the deteriorated structure after retrofitting) or loss of performance (with respect to the original structure before retrofitting) can be evaluated by analyzing the recorded strain and deflection [346].

Load testing has been mostly used for assessing the efficacy of the rehabilitations using FRP composites. Load testing is usually performed before and after the application of externally bonded FRP composites to compare the structural performance of the original deteriorated bridge and the retrofitted bridge. Hag-elsafi et al., Ettouney and Alampalli [278, 350, 351] conducted load tests of Route 378 Bridge Over Wynantskill Creek and Church Street Bridge (see Figure 5.11) in New York. The tests were performed before and after retrofitting with FRP composite laminates to evaluate the effectiveness of the strengthening system and investigate its influence on the structural behavior of the bridge.

However, if the defects in the FRP composites are not noticeable enough to contribute to the overall load deflection response, the defect may go undetected [265]. Thus this method lacks sensitivity to finer defects and further does not have the capacity to give details about the size, location, and magnitude of the defects [279].

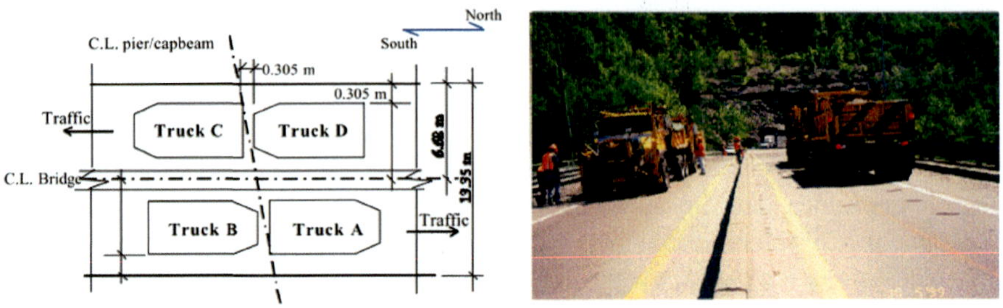


FIGURE 5.11 Load testing to assess the installation of externally bonded FRP laminates [351]

## 5.4 OTHER NDT WITH LESS RELEVANCE

### 5.4.1 Magnetic Flux Leakage Testing (MFL)

The magnetic flux leakage (MFL) technique detects the defects by measuring the leaked magnetic flux at the flaw region in ferromagnetic materials, as shown in Figure 5.12, and does not work for nonferromagnetic materials [352, 353]. However, for the internal application of FRP (nonferromagnetic material), with a potential new technique combining smart FRP bars/strips with ferromagnetic materials and the self-magnetic flux leakage (SMFL) technique, it may be possible to determine the length of the damage in addition to detecting the presence of the damages [354]. Similarly, in the case of external application of FRP, this method can be used to detect defects such as corrosion and loss of cross-section on steel reinforcements in the substrate [355].

### 5.4.2 Dye Penetrant Testing (DPT)

The basic principle behind the dye penetrant testing is to increase the contrast between the defect and the surrounding material to detect the defects that may not be evident under normal visual inspection. The step-by-step process of the dye penetrant testing is shown in Figure 5.13, which involves spraying a high penetrating dye on the FRP composites, removing the excess, and applying a developer agent that reveals the defects either by color contrast under normal visible light or by fluorescence under

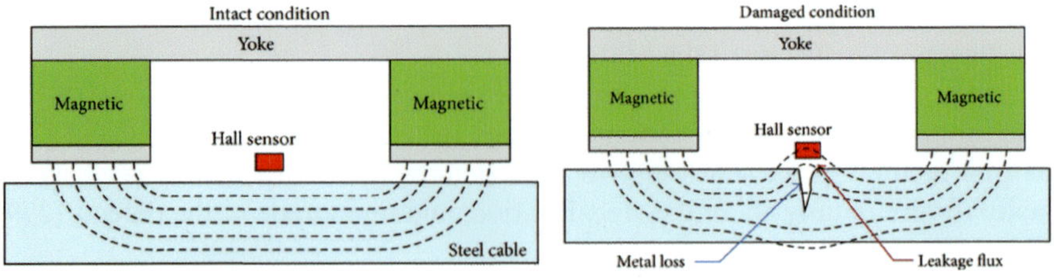


FIGURE 5.12 Magnetic flux leakage testing [356]

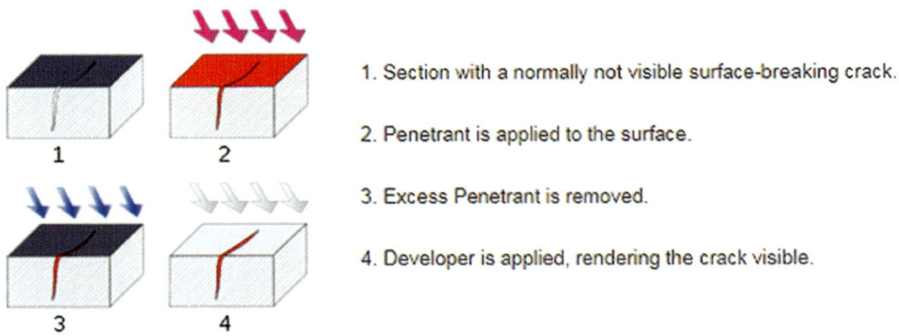


**FIGURE 5.13** Dye penetrant testing [356]

UV light [357]. This method has been mostly used for steel material; however, it could have some potential for detecting or magnifying FRP surface defects.

### 5.4.3 Eddy Current Testing (ECT)

The eddy current testing (ECT) is conducted by analyzing the variations in the flow of eddy current induced in a test object, which is caused by the alternating magnetic field generated when the alternating current is applied to an induction coil placed near that object, as shown in Figure 5.14(a). The discontinuities or flaws in the test object alter the flow of eddy current, which is detected by the pickup coil, as shown in Figure 5.14(b) [265, 282].

Gros and Takahashi [358] demonstrated that eddy current testing (ECT) can detect delamination at the interfaces between plies in carbon fiber laminates. Similarly, Mook, Lange, and Koeser [359] found that ECT can identify fiber orientation, local imperfections such as fiber fraction variations, resin-rich zones, delaminations, and impact damage in CFRP laminates. However, ECT is not suitable for electrically nonconductive materials like GFRP, but it is applicable to CFRPs due to the electrically conductive nature of carbon fibers [360, 361].

The advantages of ECT include near-surface flaw detection, rapid results, portable equipment, minimal preparation, no need for surface contact, and the ability to scan irregular shapes. Whereas disadvantages include the application only for high fiber density locations and electrically conductive materials, the need for high skills

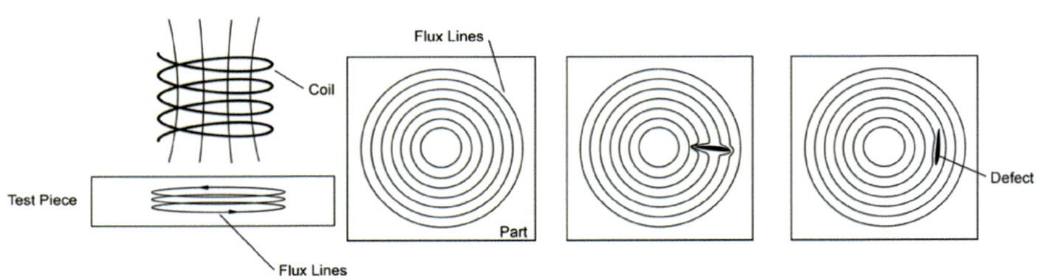


**FIGURE 5.14** Eddy current testing [265]

and training for the operator, the limited depth of penetration, and the inefficiency in detecting flaws such as delamination that runs parallel to the probe coil winding direction [279].

## 5.5 NEW TRENDS AND COMPLEMENTARY METHODS

### 5.5.1 Vision-Based Inspection

Vision-based techniques can be used at both the local and the global levels for non-destructive testing and structural health monitoring of the structures [362]. At the local level, it can be used for local condition assessment such as detection of cracks, delamination, and other surficial defects by using visual images of the surface of structures and infrared images reflecting the internal structure of the elements. Whereas at the global level, it can be used for global condition assessment by recording and analyzing global structural responses such as displacement, vibration, and modal parameters using visual tracking.

While inspecting a structure for localized defects using conventional visual inspection methods, photos or videos are used just as a reference for documenting purposes, and it is a time-consuming process demanding a considerable number of inspectors. But in computer aided vision-based techniques, the photos and videos taken are the main means of analysis and assessment. The computer-aided vision-based techniques implement image processing methods and learning-based methods or artificial intelligence (such as machine learning and deep learning) which makes condition assessment more effective in automatic ways. Qiu [363] conducted a state-of-the-art review of vision-based NDT techniques for detection of defects on civil infrastructures with bonded FRP applications. In this study, he presented and discussed the computer-aided motion magnification techniques for defect identification and recommended its use with artificial intelligence to make defect evaluation of FRP-bonded civil infrastructures highly efficient and automated. These vision-based techniques can reveal defect regions that would have been difficult to perceive by the naked eye.

On the other hand, at the global level for vision-based displacement measurements of structures under external loads, a still camera is used to track the motion of the structure. The main principle behind using vision-based monitoring is that the motion measured in pixels in an image would be converted into a physical quantity such as millimeter [364]. Figure 5.15 shows the displacement time history of point P1 in a footbridge induced due to the passing of a golf cart which is obtained by using the vision-based technique. The displacement obtained can be further processed to obtain key indicators that can be used to evaluate structural performance such as the global structural behavior, load carrying capacity (during load tests under known loads), structural dynamic properties, structural modal parameters, and levels of vibration (for serviceability requirements) and even tension force in cable (as it is related to vibration frequencies and deformations) [362].

The advantages of vision-based techniques include non-contact measurements, remote data collection, fast, low cost, less labor intensive, and minimum interference

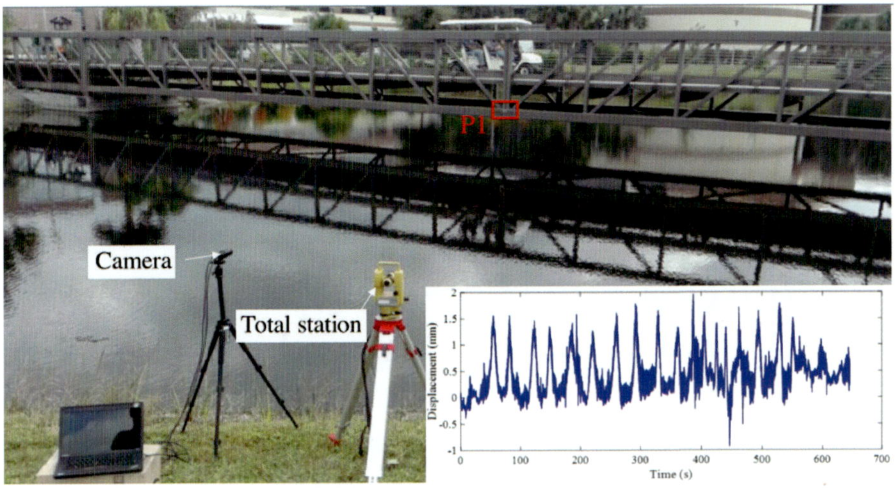


**FIGURE 5.15** Vision-based displacement measurement of a foot bridge [364]

to the daily traffic [362]. Similarly, by using processing software, the real-time results can be obtained [311]. However, its precision can be affected due to movements of the camera and heat waves that distort the field of view (due to heating of the airmass in the field of view) [365]. Other factors such as fog, rain, and sun reflection can affect the accuracy of vision-based measurements.

### 5.5.2 Automated/Robotic Inspection

Current bridge inspection practices are labor-intensive and mostly conducted manually, which is a time-consuming process not adequate and efficient enough for the increasing demand for inspection and maintenance of bridges [366–368]. Moreover, manual inspections are not always safe and cannot access hard-to-reach and confined spaces. In contrast to all these limitations of manual inspection, the use of technologically advanced semi or fully autonomous robots for NDT applications provides the advantages of efficiency, low safety risk, and low cost. It requires relatively less time for the collection of data. Hence, in recent years, there has been considerable inclination toward the use of NDT techniques integrated with robots.

The robots used for NDT applications can be divided into aerial robots, ground robots, and underwater robots [366], as shown in Figure 5.16. Aerial robots include Unmanned Aerial Vehicles (UAVs) (commonly referred to as drones), ground robots include climbing robots, wheeled robots, sliding robots, and legged robots, and underwater robots include unmanned submersible vehicles (USVs), underwater marine vehicles (UMVs), and underwater vehicles (UUVs). All these wide arrays of diverse robots have been developed with the main purpose of making the bridge inspection less time-consuming, safe, and relatively cost-effective.

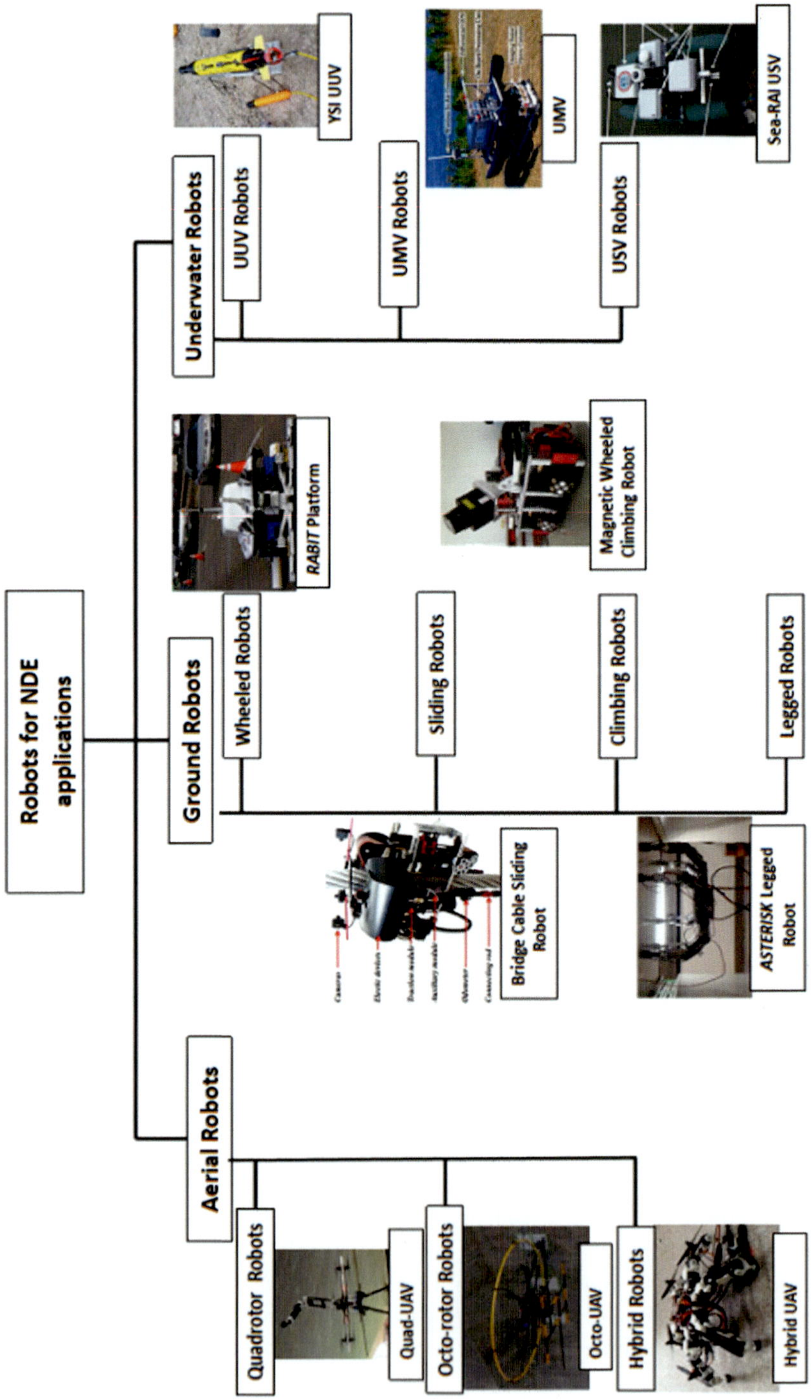


**FIGURE 5.16** Different categories of robots used for NDT applications [366]

However, there are some limitations to the use of robotic inspection. Environmental factors such as improper lighting, ambient noise, and high wind speed affect the operation of robotic inspection. Factors such as restrictive regulations for the operation of drones and challenging congested areas for flight operations are the main limitations in using UAVs [369, 370]. Similarly, underwater robots that use vision-based sensors would be influenced by the clarity of the water and weather conditions [366].

## 5.6 NEW METHODS FOR IMPROVING THE DETECTABILITY OF FRP BARS IN CONCRETE

Khedmatgozar Dolati et al. [371] suggested new methods for the enhancement of the detectability of Fiber-Reinforced Polymer (FRP) bars embedded in concrete using modifications to the bar surface. They used Ground-penetrating radar (GPR) and Phased Array Ultrasonic (PAU) testing to assess the detectability of both commercially available and modified FRP bars. The modifications, including the coating of metallic particles and wire winding on the surface of the FRP bars, were introduced to improve the response of these bars to NDT methods. The results of their study indicated that modified FRP bars demonstrated significantly improved detectability compared to unmodified bars. Specifically, the modified bars were detectable at greater depths and with greater clarity, allowing for more accurate identification of embedded FRP bars and their potential defects. The study highlights the effectiveness of these modifications in enhancing NDT capabilities and suggests their potential for broader application in civil engineering projects. Development of the new methods and their verification are discussed by Khedmatgozar Dolati et al. and Malla et al. [371, 372].

Figure 5.17 shows the schematic diagram of the alterations introduced in FRP bars. Figure 5.17(a) represents all the various types of FRP bars available based on the methods used to establish bond between the rebars and the concrete by creating variations in the geometry of their cross section and surface including but not limited to helically wrapped (HW), sand coated (SC), helically wrapped sand coated (HWSC), indented (In) or grooved, ribbed (Rb), and other FRP bars.

The concept and principle behind the proposed modifications is to provide metallic presence on the FRP bars to make them detectable by existing NDT methods that are tuned to detect metallic reinforcing bars. The methods for making embedded FRP detectable by NDT could be the method of choice for ensuring their inspectability and provide the following advantages:

1. It is easy to implement; the FRP surface modification can be performed during FRP bar fabrication or later using similar or the same machinery to what is currently being used,
2. Requires no modifications to the NDT devices readily available for customary reinforced concrete,
3. Requires no special training for the inspectors,

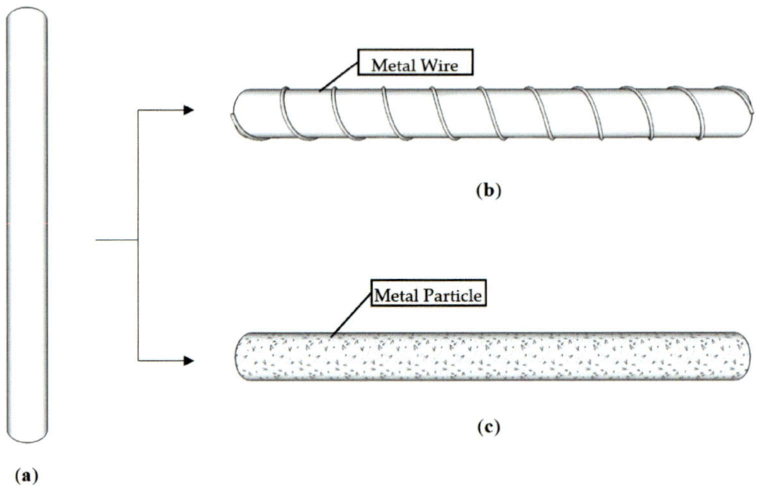


FIGURE 5.17 Modification of FRP Bar: (a) FRP bar; (b) FRP bar with metal wire; (c) FRP bar with metal particle coating (with or without bond enhancement coating)

4. It is durable and corrosion resistant as the metallic coating or wire would be impregnated within resin,
5. The added cost is not significant and
6. It will eliminate the uncertainty in the use of FRP due to the present limitations on their inspection.

### 5.6.1 Validation of New Methods

For validation of new methods, several panel specimens similar to those in Figure 5.18 were cast using concrete with a nominal compressive strength of 4600 psi (31.7 MPa), in accordance with ASTM C39 [373]. The modified bars also had discontinuities to simulate the occurrence of damages in the FRP bars and to check the capability of the NDT devices in detecting them.

It was successfully demonstrated that the detectability of FRP bar in concrete can be improved through the application of metallic particle coating and wire winding modifications. Both GPR and PAU testing methods showed enhanced detection capabilities with these modifications, particularly at greater depths and with increased sensitivity to damage. The findings revealed that while traditional GPR and PAU methods struggle to detect unmodified FRP bars, the proposed modifications significantly enhanced their detection in both line views and depth slices. Figure 5.19 shows an example of the detection of bars and defects by PAU. Without the modifications, PAU has little or no detection capability. The improved detectability of modified FRP bars has the potential to influence future standards and practices in the inspection of FRP-reinforced concrete, promoting the broader adoption of FRP materials in civil engineering [371, 372].

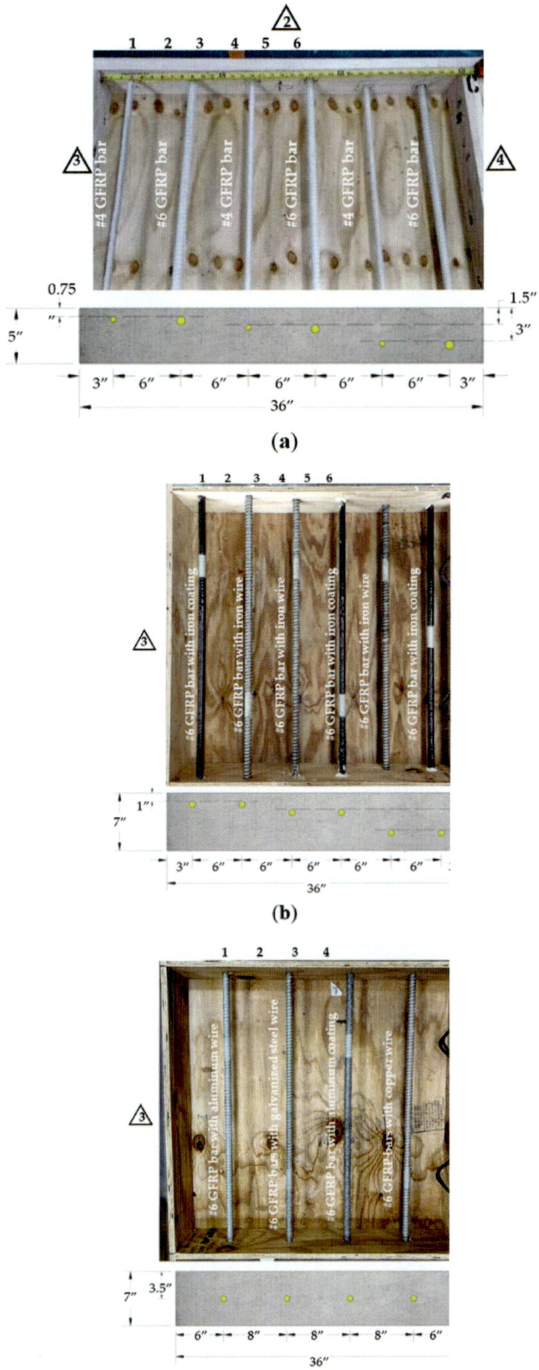


**FIGURE 5.18** Reinforcement/Dimension details of detectability slab specimens: (a) Slab A; (b) Slab B; (c) Slab C

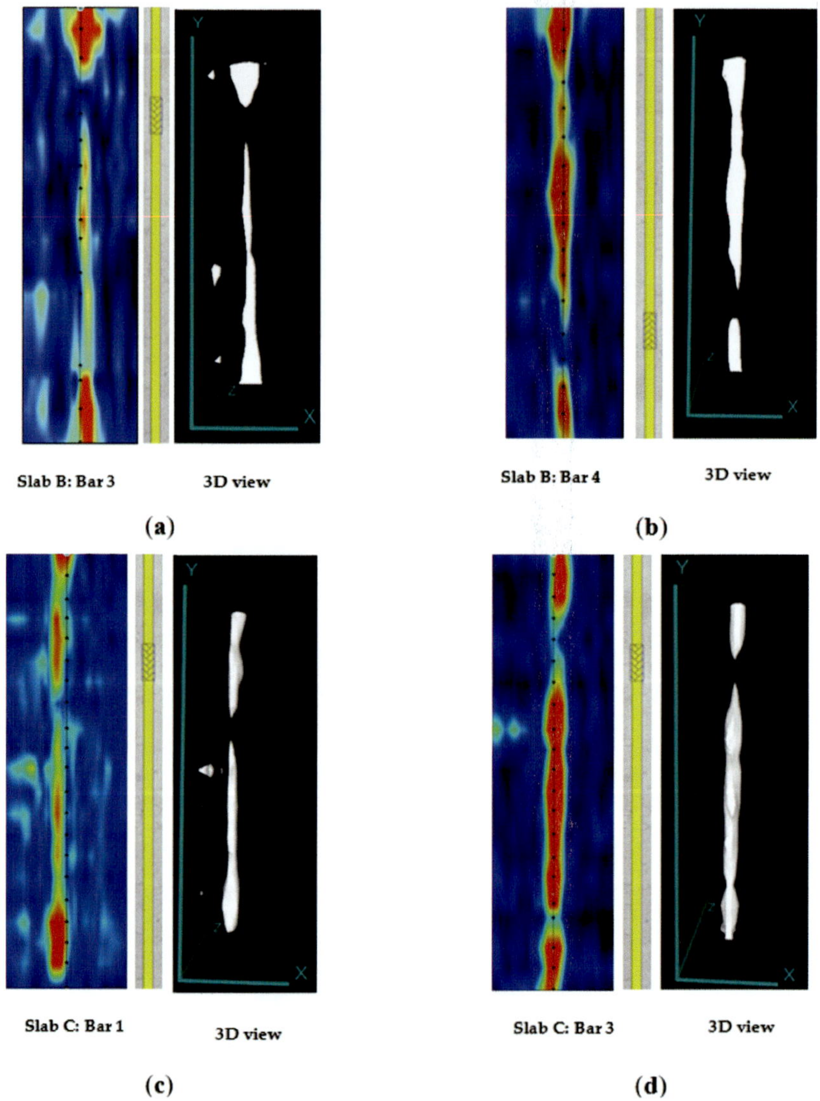


**FIGURE 5.19** Depth slices of slabs for detectability of modified GFRP bars

## 5.7 SUMMARY OF ADVANTAGES AND DISADVANTAGES OF NDT METHODS APPLICABLE FOR DAMAGE/DEFECT DETECTION OF FRP-RSC

All the findings from the literature survey conducted serve to assess the suitability of each individual method toward in-situ inspections of FRP-RSC elements. The advantages and disadvantages of the individual methods discussed in the previous sections have been presented in tabulated form (see Table 5.2) to provide the basis

## TABLE 5.2
**Advantages and disadvantages of NDTs applicable for FRP inspection**

<table>
<tr><th>Method</th><th>Advantages</th><th>Disadvantages</th></tr>
<tr><td>Visual Inspection (VT)
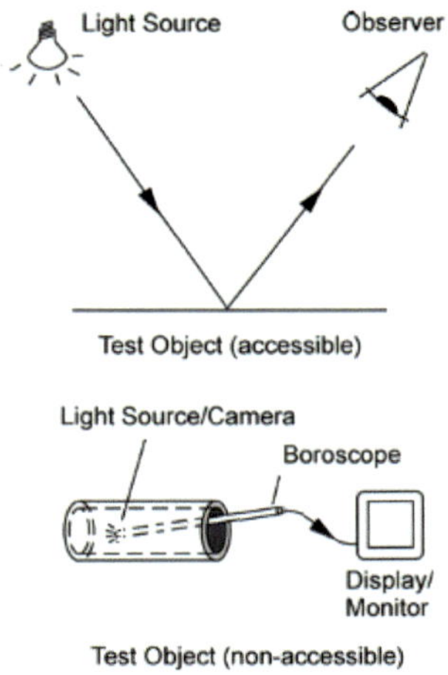
</td><td>

- Checks obvious and outstanding anomalies on a global basis
- Fast, simple, and cost-effective
- Real-time inspection and can be interpreted instantly
- Sets the baseline for other NDT methods
- Based on the findings of visual inspection decisions can be made on necessity of further inspection with more sophisticated NDT methods
- Can be performed before any further NDT as a supplementary method

</td><td>

- Detects only surface defects
- Subjective and observations may vary depending on visual perceptions of individuals

</td></tr>
<tr><td>Tap Testing (TT)
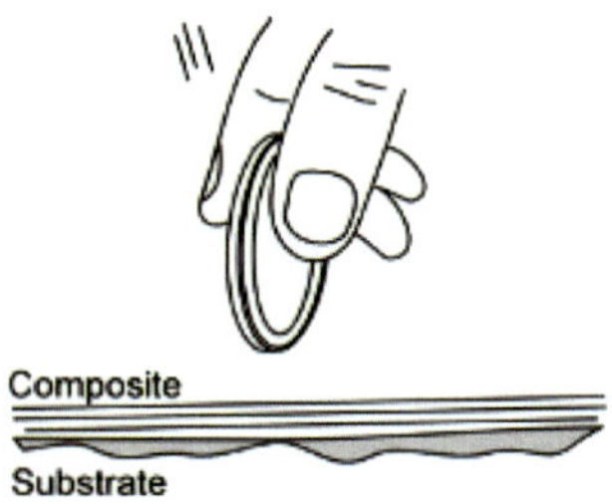
</td><td>

- Inspections can be done faster
- Can inspect large areas
- Low cost
- Easy to use method that requires less training to implement
- Real time and provides immediate results

</td><td>

- Subjective as it relies on the human factors
- Results vary due to the inconsistencies in the force, angle and devices used
- Ambient noise and geometric changes may lead to erroneous interpretation
- Does not provide the depth and width of the defects

</td></tr>
<tr><td>Impact Echo (IE)
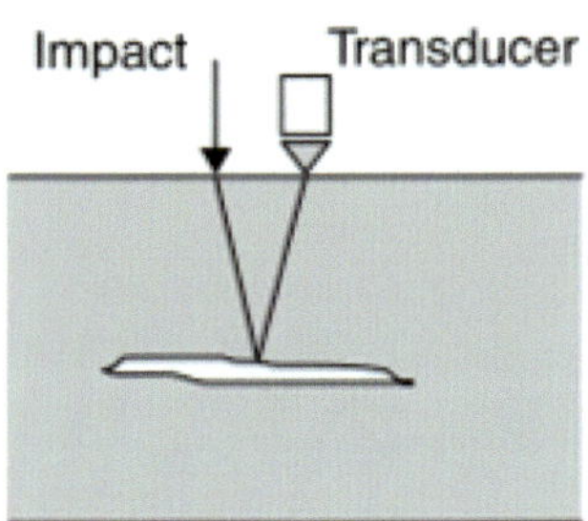
</td><td>

- Has deep penetrating capability into concrete
- Promising for the evaluation of cracks, voids, delamination, and discontinuities in concrete
- Unlike ultrasonic methods, impact echo utilizes lower frequency which allows the impact echo to overcome high-signal attenuation encountered with ultrasonic methods
- It needs only one surface to be accessible for testing

</td><td>

- Current instrument limited to testing members 40 in. (1 m) thick
- Requires experienced personnel to operate
- Smaller cracks and discontinuities cannot be detected easily

</td></tr>
</table>

*(Contiued)*

## TABLE 5.2 (CONTINUED)
**Advantages and disadvantages of NDTs applicable for FRP inspection**

| Method | Advantages | Disadvantages |
|---|---|---|
| Ground-Penetrating Radar (GPR)<br>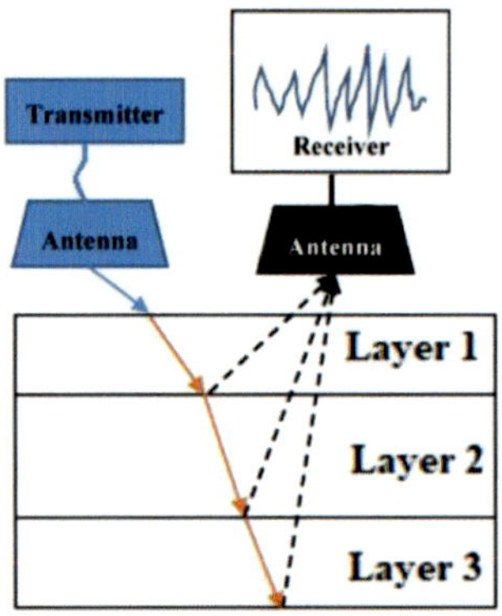 | • Penetrates beyond concrete–air interfaces which allows inspection of features below the interface<br>• Detects defects at a greater depth than other NDT techniques such as infrared thermography<br>• Air-coupled GPR allow rapid inspection | • Relatively more expensive than other NDT methods<br>• Not suitable for detecting air-filled defects |
| Ultrasonic Testing (UT)<br>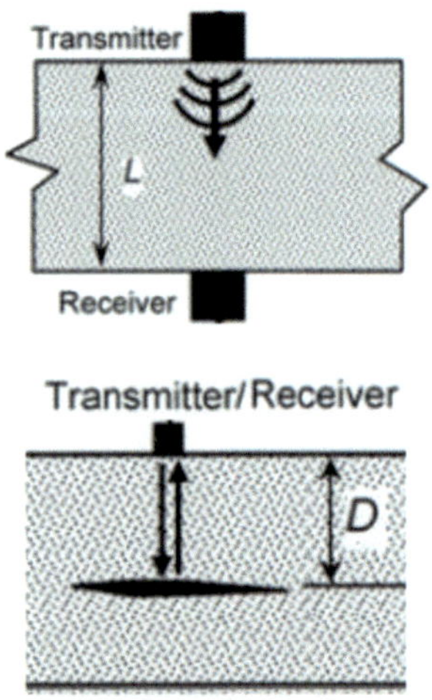 | • Fast, easy to use in fields, and has good resolution<br>• Can penetrate well into the materials and detect subsurface flaws<br>• Detects wide range of defects such as debonding, resin variations, broken fibers, impact damage, moisture, cracks, voids, and subsurface defects | • Requires highly trained individual to conduct the test and interpret the data<br>• UPV requires access to two sides of the member<br>• Ultrasonic echo method is applicable to a limited thickness of members |
| Infrared Thermography (IR)<br>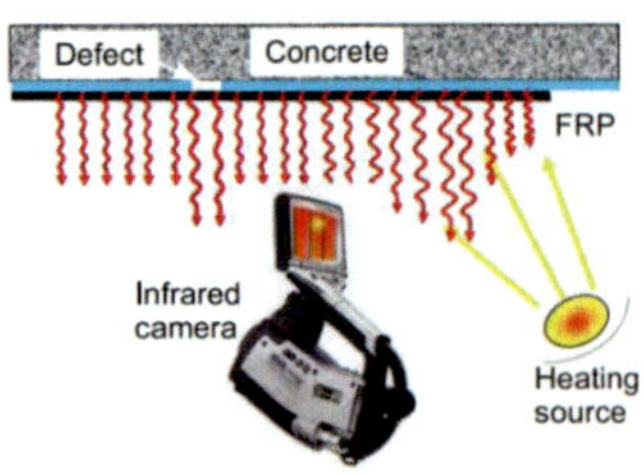 | • Detects subsurface defects globally over larger surface areas<br>• Fast and inexpensive<br>• Inspections and data interpretation can be done in real time | • Not that reliable in detecting water-filled defects<br>• Cannot detect defects present deep within the concrete effectively<br>• Requires proper environmental conditions |

*(Contiued)*

## TABLE 5.2 (CONTINUED)

**Advantages and disadvantages of NDTs applicable for FRP inspection**

| Method | Advantages | Disadvantages |
|---|---|---|
| Acoustic Emission (AE)<br>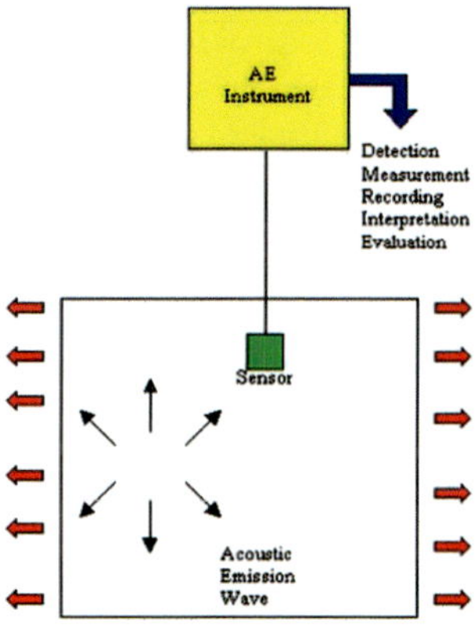<br> | • Allows fast and global inspections<br>• Can be performed without shutting down the service of the structure<br>• Detects defects due to fatigue loading such as fatigue cracks, fiber fractures, matrix micro-cracks, fiber-matrix debonding, and delamination<br>• Effective in locating the origin of discontinuities<br>• Detects the growth rate of a defect | • Difficulty in differentiating damage source<br>• Requires high level of skill in correlating the AE data to respective damage sources<br>• Noisy environment may lead to false signals<br>• Defects might go undetected if the loading is not sufficient<br>• Cannot give information on the type and size of defects |
| Laser Interferometry<br>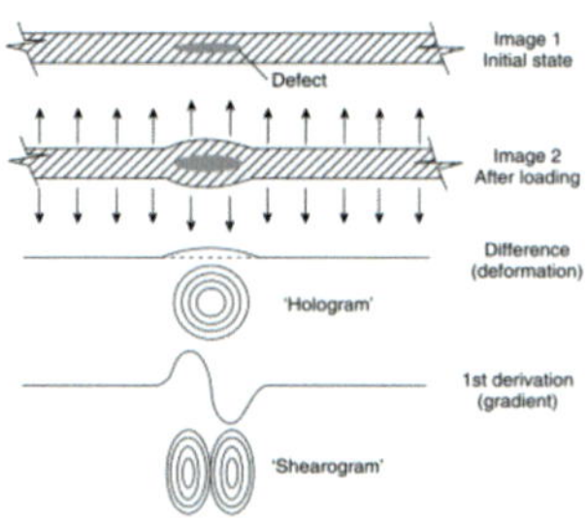<br> | • Non-contact method<br>• High sensitivity<br>• High detection rate | • Require laborious analysis of the results<br>• Expensive device |
| Acoustic-Laser Vibrometry<br>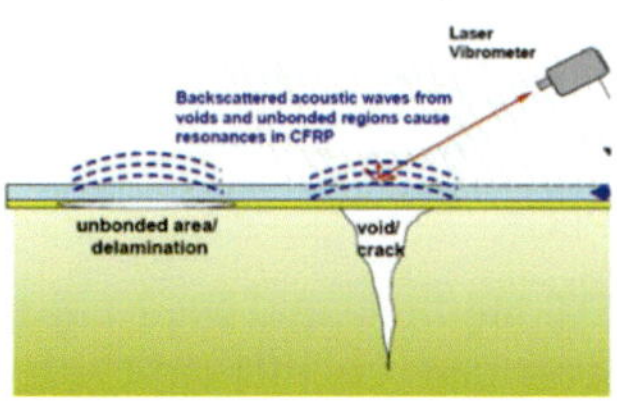<br> | • Non-contact method<br>• Less harmful to people<br>• Capable to detect small-scale defects | • Hearing impairment to people<br>• Eyes protection is needed<br>• Noise may affect the effective signals |
| Global Structural Response Testing (GSR)<br>• **Modal Testing**<br>• **Load Testing**<br> | • Used for assessing the efficacy of the FRP rehabilitations<br>• Used for long-term performance of structures | • The local defects may not be significant enough to contribute to global response<br>• Lacks sensitivity to finer localized defects and cannot give information on their size, location, and magnitude |

*(Contiued)*

**TABLE 5.2 (CONTINUED)**
**Advantages and disadvantages of NDTs applicable for FRP inspection**

| Method | Advantages | Disadvantages |
|---|---|---|
| Vision-based Techniques 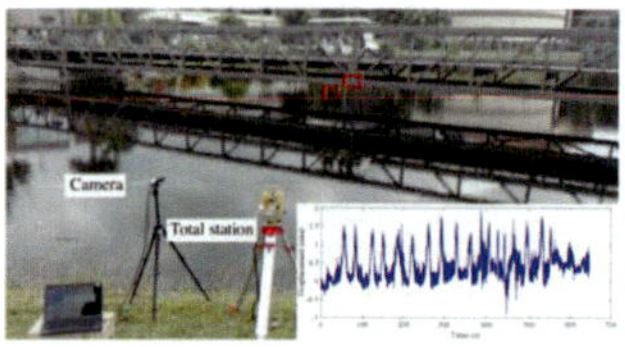  | • Non-contact measurements<br>• Remote data collection<br>• Low cost and less time consuming<br>• Less labor-intensive<br>• Minimum interference to traffic | • Precision can be affected by vibrations and movement of camera<br>• The field of view could be distorted due to environmental factors which give erroneous observations |
| Robotic Inspection  | • Less time consuming<br>• Automated robotic inspections are quite efficient<br>• Can reach places where humans cannot<br>• Reduces the risk of accidents involving humans | • Environmental factors such as improper lighting, ambient noise, high wind speed, weather conditions, low visibility, etc. affect vision-based robotic inspection<br>• Restrictive regulations in flying UAVs<br>• Difficulty in maneuvering robots in congested space |

for the selection of the methods most applicable. The presentation of the findings in tabular format allows easier comparison of methods and a clear understanding of the relative advantages and disadvantages of each individual method.

## 5.8 PROMISING/POTENTIAL NDT METHODS FOR DAMAGE/DEFECT DETECTION IN FRP-RSC ELEMENTS

Analysis of the information collected about various NDT methods and their capabilities in relation to the detection of damages in general and for FRP-RSC elements specifically can lead to the identification of promising or potential methods for further investigation. Aside from referring to experiences and expert opinions, a statistical overview of the prior relevant investigations can prove to be very helpful and reliable. For statistical analysis, first, the damages/defects in FRP-RSC elements need to be broadly categorized into defect groups that have similar features as they concern the application of NDT methods. As shown in Table 5.3 (similar to Chapter 4), the NDT methods for FRP inspection are grouped for FRP and concrete in the internal and external applications. For external applications, defects in the FRP can be generalized as bond defects, defects within FRP, and surface anomalies, and

**TABLE 5.3**
**Damage categories for statistical summary of the applicable NDT methods for FRP inspection**

| External application of FRP | | Internal application of FRP | |
|---|---|---|---|
| **Defects in FRP** | **Defects in concrete substrate** | **Defects in FRP** | **Defects in concrete** |
| • Bond defects<br>• Defects within FRP composite layers<br>• Surface anomalies | • Cracks in concrete<br>• Voids in concrete<br>• Delamination in concrete<br>• Corrosion in steel reinforcement | • Defects at the interface of FRP-concrete<br>• Defects in FRP reinforcing bars, strands | • Cracks in concrete<br>• Voids in concrete<br>• Delamination in concrete |

defects in concrete can be categorized as cracks, voids, and delamination in concrete and corrosion of steel reinforcing bars. For internal applications, defects in the FRP can be generalized as defects at the interface of FRP and concrete (bond defects) and defects within FRP (bars or strands), and defects in concrete can be categorized as cracks, voids, and delamination in concrete.

### 5.8.1 Statistical Summary of NDT Methods for External Application of FRP

The findings of a comprehensive literature survey of 100 past studies on the application of NDT methods in detecting damages in the external application of FRP are shown in Table 5.4 (for damages in FRP), and its statistical summary is shown in Table 5.5 and Figure 5.20.

Similarly, the statistical summary of the NDT methods applicable for detecting defects in concrete has been provided from a study by the authors [298] are shown in Figure 5.21.

### 5.8.2 Statistical Summary of NDT Methods for Internal Application of FRP

As mentioned earlier, literature on the application of NDT methods for the internal application of FRP is limited and scarce. This represents a research gap that this research study attempts to address as much as possible. Hence this study can be used as an initial step for propelling future studies on nondestructive testing for internal application of FRP.

Ékes [435] demonstrated for the first time that GPR can detect both CFRP and GFRP bars embedded in concrete and therefore concluded that it is a suitable tool for locating FRP bars on bridge decks. In this study, conducted in the Materials Laboratory at the University of British Columbia (UBC), two test blocks embedded with GFRP and CFRP rebars were tested with high-frequency GPR units. The test results can be seen as in Figure 5.22, where the hyperbolas indicate the presence of rebars embedded in concrete. However, this study does not give any information about the damage detectability of the GPR testing, let alone the level of its accuracy.

W. Li et al. [276] investigated the debonding behavior between CFRP rebars and concrete using AE technique and determined that the AE responses can be correlated with the damage evolution during the debonding. In this study, the CFRP rebars embedded in concrete specimens were subjected to pullout tests and the damage evolution was closely monitored using AE system. This study found that the AE parameters, such as AE hits, amplitude, and peak frequency, can be correlated to the mechanism of damage evolution, such as initiations of microstructural damage, loss of chemical adhesion, breakage of the CFRP helical layer, mechanical interlocking-related damage, fiber–matrix debonding within the CFRP rebar, and fiber breakage within the CFRP rebar. Similarly, in another study, Ghaib [436] concluded that AE signals showed correlation with different damage mechanisms of CFRP and GFRP

## TABLE 5.4

## Past studies on the application of NDT methods in detecting damages in FRP for external application

| Damages in FRP | NDT Methods | Damage Detectability | References |
|---|---|---|---|
| Bond defects (debonding and voids between FRP and concrete substrate) | TT | Detects debonding as small as 1.5 in × 0.75 in (1.1 in$^2$) | [284] |
| | | Qualitative detection of debonding | [283–285, 374, 375] |
| | | Qualitative detection of voids | [283–285] |
| | IE | Detects debonding at the epoxy–concrete interfaces of NSM CFRPs. | [376] |
| | | A mobile impact-echo device can detect debonding rapidly and evaluate large areas (long lengths) of CFRP plates | [377] |
| | | Qualitative detection of voids | [291] |
| | | Can potentially provide bridge engineers a means to rapidly evaluate long lengths of CFRP laminates applied external on bridges | [378] |
| | MW | Provides a close estimate of the spatial dimensions of debonding regions. | [379] |
| | | Detects debonding regions as small as 2 cm × 0.5 cm and provide important dimensional information such as spatial extent, location, and the geometry of a debonding region | [302] |
| | | Detects size and severity of debonding | [303] |
| | | Detects size and severity of voids | [303] |
| | | Qualitative detection of air voids | [306, 380] |
| | | Determination of spatial dimensions and locations of debonding/ delaminated areas | [381] |
| | | Detects location and size of the voids and debonding, it can further detect their depth | [382] |
| | | Qualitative detection of debonding | [374] |
| | GPR | Higher frequency antenna of about 2 GHz can easily detect debonding (but lower frequency of 1 GHz cannot) | [313] |
| | | 1.5 GHz ground-coupled GPR antenna performed typically well in detecting water-filled defects/voids (as small as 50.8 × 50.8 × 1.5 mm$^3$) but its accuracy was not that satisfactory in detecting air-filled defects/ voids | [308, 314] |
| | | Qualitative detection of voids | [315] |
| | | Far-field airborne radar can detect debonding in the GFRP–concrete interface regions | [383] |
| | | Qualitative detection of bond defects | [140] |
| | | Determines the in-service bond conditions of externally applied FRP laminate | [384] |
| | UT | Qualitative detection of debonding | [25, 292, 317–319, 375, 385–390] |
| | | Qualitative detection of voids | [25, 292] |
| | | Detects debonds of 0.25 inch in diameter | [391] |
| | | Qualitative detection of bond defects | [392] |
| | IR | Qualitative detection of debonding | [140, 313, 315, 328, 375, 388, 393 –407] |
| | | Quantitative detection of debonding | [339, 408] |
| | | Qualitative detection of voids | [315, 329, 402] |
| | | Quantitative detection of voids | [409] |
| | | Detects air-filled debonding of sizes 3 in × 3 in, 2 in × 2 in, and 1.4 in × 1.4 in along with water-filled debonding in both GFRP and CFRP applications | [410] |
| | | Provides qualitative and quantitative information on bond defects between FRP and concrete substrates | [411] |
| | | Qualitative detection along with the assessment of the severity of damage of composite–concrete interface debonding | [412] |

*(Contiued)*

## TABLE 5.4 (CONTINUED)

## Past studies on the application of NDT methods in detecting damages in FRP for external application

| Damages in FRP | NDT Methods | Damage Detectability | References |
|---|---|---|---|
| | | Measures the size of different bond defects in CFRP | [413] |
| | | Depth detection of bond defects | [414] |
| | | Identifies delamination regions/ voids that are at least 5 cm × 5 cm in size | [415] |
| | | Detects near-surface voids (<10 cm from surface) | [416] |
| | | Simultaneous use of thermal shearography and thermography for defect identification | [417] |
| | AE | Detect and analyze the progression of crack leading to FRP debonding | [333] |
| | | Differentiate between concrete cracking and CFRP debonding | [334] |
| | | Qualitative detection of debonding | [375, 396] |
| | LT | Quantitative detection of voids | [409] |
| | | Qualitative detection of voids | [375] |
| | | Qualitative detection of debonding | [338, 339, 375, 408, 418] |
| | | Detects and quantifies the size of bond defects | [341, 342, 419] |
| | RT | Qualitative detection of debonding | [12, 375] |
| | | Qualitative detection of voids | [12, 375] |
| Defects within FRP composites layers (debonding, voids, delamination at composite layers) | TT | Qualitative detection of delamination | [283–285, 287, 288, 375] |
| | | Acoustic Impact Testing which is analogous to coin tap testing is very sensitive in identifying delamination in the range 0.01 to 0.25 in$^2$ (0.254 to 160 mm$^2$) | [420] |
| | IE | Capable of testing delamination in members of thickness up to 40 in. (1 m) | [289] |
| | | Qualitative detection of delamination | [291, 293] |
| | MW | Delamination smaller than 20 mm × 20 mm | [307] |
| | | Qualitative detection of delamination | [421, 422] |
| | GPR | Higher frequency antenna of about 2 GHz can easily detect delamination (but lower frequency of 1 GHz cannot) | [313] |
| | UT | Qualitative detection of delamination | [316, 375, 386, 421, 423–429] |
| | | PAU identifies both the size and location of delamination | [324–326] |
| | | PAU detects size and location of defects in CFRP and GFRP | [322] |
| | IR | Qualitative detection of delamination | [313, 361, 375, 396, 397, 404–406, 421, 423, 426] |
| | | Pulse phase thermography (PPT) determines the layer at which a defect is present | [414] |
| | | Qualitative detection along with the assessment of the severity of damage of interlaminar debonding within the composite | [412] |
| | | Detectivity of delamination using lock-in thermography is satisfactory whose sensitivity depends on defect's depth, decreasing linearly with depth | [430] |
| | | Lock-in thermography identifies both the size and location of delamination | [325] |
| | AE | Qualitative detection of delamination | [396, 431] |
| | LT | Qualitative detection of delamination | [340, 396] |
| | | Quantitative evaluation of defects is possible with the measurement of their width and depth | [432] |
| | | Simultaneous use of thermal shearography and thermography for defect identification | [417] |
| | RT | Qualitative detection of delamination | [12, 346, 375, 424] |
| | | Qualitative detection of voids | [426, 433, 434] |
| FRP surface anomalies (wrinkling, blistering, scratches, discoloration, cracks) | VT | Visual testing can be used for qualitative and quantitative detection of almost all surface anomalies | |

*Note:* VT = Visual Inspection, TT = Tap Testing, IE = Impact Echo Testing, MW = Microwave Testing, GPR = Ground-Penetrating Radar, UT = Ultrasonic Testing, IR = Infrared Thermography Testing, AE = Acoustic Emission Testing, LT = Laser Testing, RT = Radiographic Testing

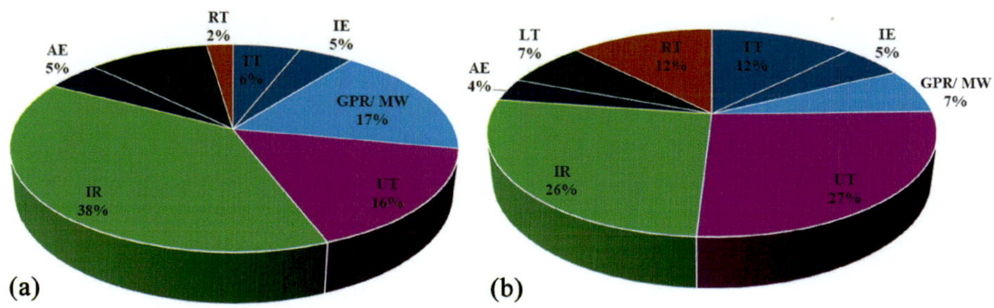


**FIGURE 5.20** Statistical summary of NDT applicable for (a) bond defects and (b) defects within FRP composite layer

## TABLE 5.5
## Statistical summary of Table 5.4

| NDT<br>Damages in FRP | TT | IE | MW/<br>GPR | UT | IR | AE | LT | RT |
|---|---|---|---|---|---|---|---|---|
| Bond defects | 5<br>(6%) | 4<br>(5%) | 15<br>(17%) | 14<br>(16%) | 33<br>(38%) | 4<br>(5%) | 9<br>(11%) | 2<br>(2%) |
| Defects within FRP composite layers | 7<br>(12%) | 3<br>(5%) | 4<br>(7%) | 15<br>(27%) | 15<br>(26%) | 2<br>(4%) | 4<br>(7%) | 7<br>(12%) |
| FRP surface anomalies | Use VT | | | | | | | |

bars subjected to tensile load. Hence, AE testing has been used in the past for detecting debonding between the embedded FRP rebars and surrounding concrete and has good potential as a nondestructive testing method applicable to embedded FRP rebars. Nevertheless, AE will not be able to detect damage/defects that are already present in the concrete elements.

Based on the review of available literature and the experience of the authors, the following nondestructive devices are identified to be evaluated as potential methods for the detection of damages (defects in the interface and defects in FRP reinforcing bars, strands) in the internal application of FRP in the following chapters:

- Ultrasonic Testing (Phased Array Ultrasonic)
- Ground-Penetrating Radar
- Infrared Thermography
- Impact Echo

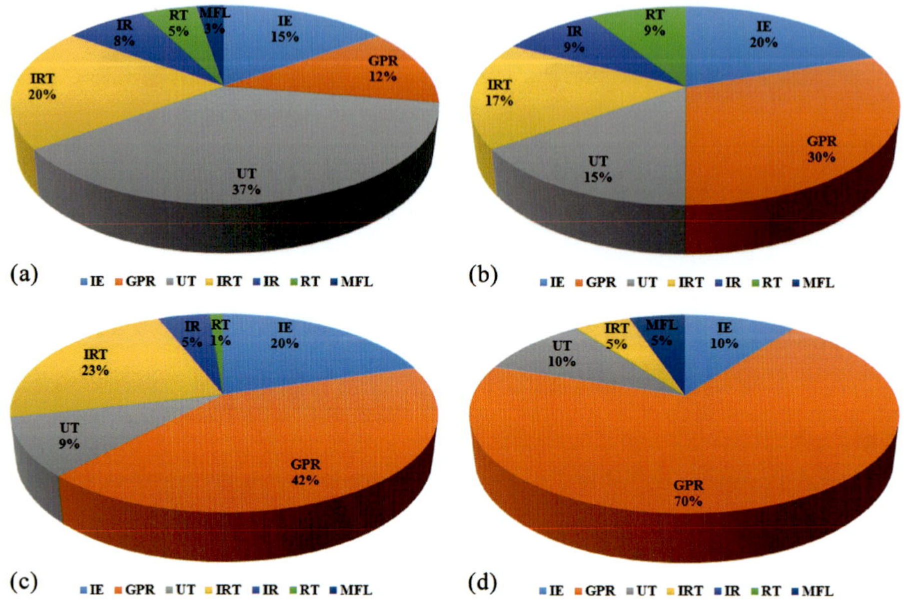


**FIGURE 5.21** Statistical summary of NDT applicable for (a) cracks in concrete, (b) voids in concrete, (c) delamination in concrete, and (d) corrosion in steel reinforcement

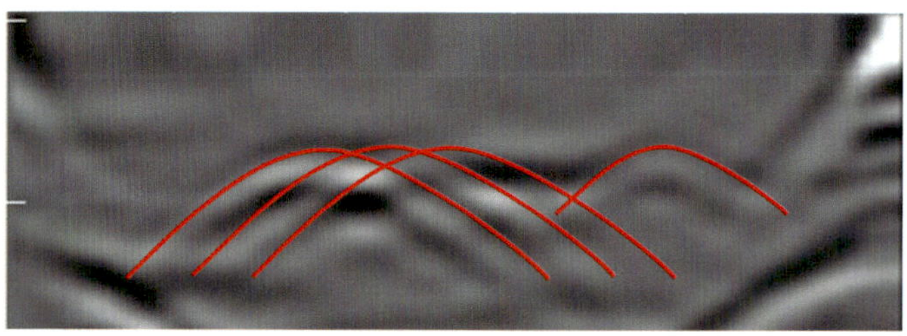

**FIGURE 5.22** GPR scans of test slabs embedded with FRP rebars

Further, it should be noted that, for defects in concrete in the internal application of FRP, the findings shown in Figure 5.21(a)–(c) will be applicable.

Refer Chapter 6 for the selection of NDT methods and their order of priority based on FRP application and damage type.

# 6 Inspection Framework for FRP-RSC Bridge Elements

## 6.1 INTRODUCTION

### 6.1.1 Purpose of the Guide Framework

While inspection methods and codification of deficiencies have been in use for a long time for conventional steel and reinforced concrete bridges, a standardized and unified methodology or framework for inspection and damage detection of concrete bridge elements reinforced or strengthened with fiber-reinforced polymer (FRP) does not exist. The lack of clear guidelines and effective methods for condition assessment of FRP-reinforced or -strengthened concrete (FRP-RSC) elements could have negatively affected the proliferation of its use. Availability of such means and methods will have positive effects in increasing the use of FRP-RSC.

The inspection framework described in this chapter is based on the findings of research performed at Florida International University [437]. This Framework for Field Inspection of In-service FRP-Reinforced/Strengthened Concrete Bridge Elements contains the information available to date from other sources and work performed by the authors. It also represents the foundation of a framework for inspection and condition assessment of in-service FRP-RSC bridge elements that can be built upon new information that will become available in the future. Other chapters in this book can be referred to for more details on each subject.

### 6.1.2 Applicability of the Guide

The framework can be readily used by inspectors and practitioners and alleviates the concerns of bridge owners regarding uncertainties associated with the condition assessment of bridges using FRP and their maintenance. Although FRP has been used with different materials such as steel, timber, and masonry in bridge structures, this framework will specifically focus on the use of FRP with concrete bridge elements. The primary impact of this framework is to ensure the safety and integrity of bridges using FRP. The development of this framework serves as an effective catalyst to take advantage of the many benefits of FRP for the construction of bridges.

DOI: 10.1201/9781003646969-6

### 6.1.3 Intended Users

This framework targets bridge engineering professionals and inspectors who inspect, assess, or evaluate FRP-RSC bridge elements. This framework, along with the rest of this book, provides a comprehensive background on FRP composites, compares design and installation/construction with respect to the conventional reinforced concrete (RC) elements, recognizes and classifies various deficiencies observed in FRP-RSC elements, and identifies non-destructive methods for the inspection of FRP-RSC elements.

### 6.1.4 Organization of the Guide Framework and How to Use

The framework is organized into practical tables, flowcharts, and procedures to assist the inspectors in identifying and recording the damage and defects.

Section 6.2 discusses the identification and classification of observed deficiencies in FRP-RSC elements. In addition, it introduces the possible sources of those deficiencies whose presence could be an indication that deficiencies could be expected. Inspectors should familiarize themselves with this section before going to the field for inspection.

Section 6.3 presents guidance on the inspection of FRP-RSC bridge elements by explaining several types of available inspection methods, their applicability, and the inspection procedures. Inspectors should review this section to assist them in the selection of the most effective method for the condition of the bridge elements.

Sections 6.4 and 6.5 describe the type of inspection and inspector qualification.

Section 6.6 describes the procedures for inspection of FRP-RSC elements. Inspectors can follow these procedures for their inspection program.

Section 6.7 presents guidance for collection and recordkeeping of data for inspection of FRP-RSC bridge elements. It also suggests several inspection forms to facilitate the recordkeeping.

## 6.2 DAMAGES AND DEFECTS IN FRP APPLICATION

This section discusses the identification and classification of observed deficiencies in FRP-RSC elements. In addition, it introduces the possible sources of those deficiencies whose presence could be an indication that deficiencies could be expected. Two types of FRP applications are considered: External Application and Internal Application.

### 6.2.1 External Application

This refers to the application of FRP for the strengthening of concrete elements. Figure 6.1 summarizes the type of defects and damages discussed in detail in previous chapters, and Table 6.1 provides a brief description of defects and damages, their causes, and severity.

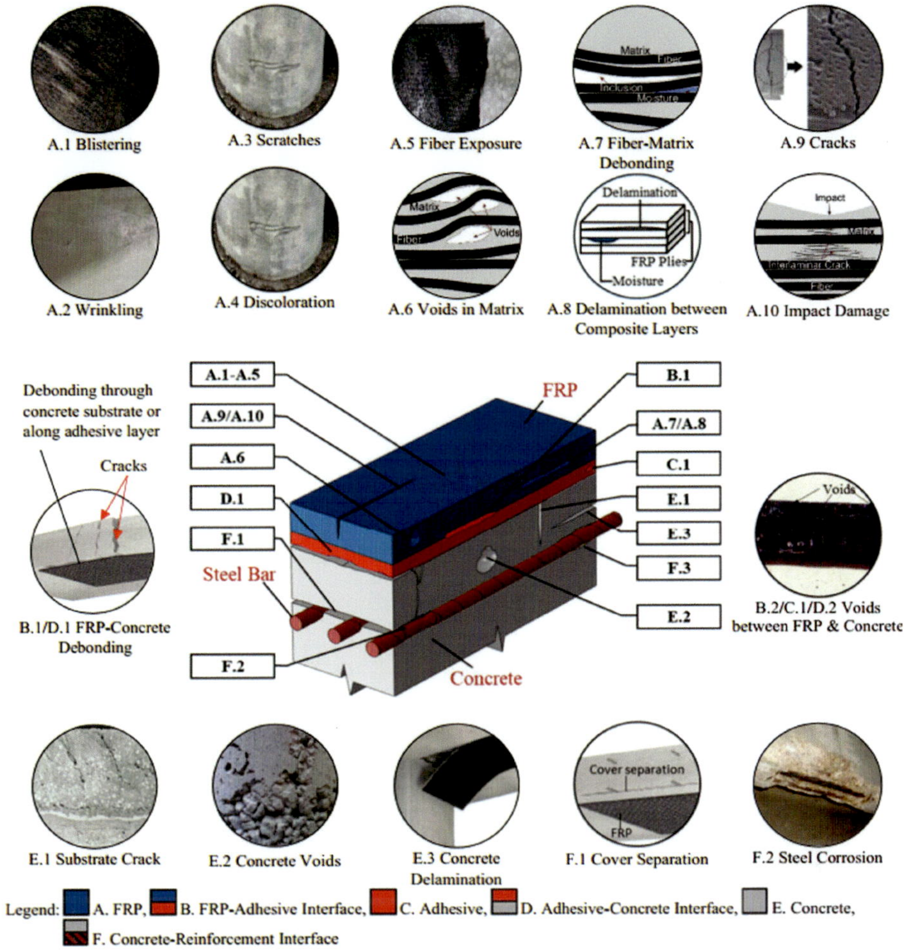


**FIGURE 6.1** Deficiencies in external FRP application

### 6.2.2 Internal Application

This refers to the application of FRP as reinforcement for concrete elements. Figure 6.2 summarizes the type of defects and damages discussed in detail in previous chapters, and Table 6.2 provides brief description of defects and damages, their causes, and severity.

## 6.3 INSPECTION

Deficiencies in FRP-RSC elements negatively affect the structural integrity and serviceability of FRP-RSC elements. NDT tools capable of detecting these defects for both the embedded FRP rebar and the externally bonded FRP composite are

**TABLE 6.1**

**Defects/damage for external application of FRP, causes, and severity**

| Defects | Description | Causes | Severity |
|---|---|---|---|
| | **Defects in FRP composite (Defect category A)** | | |
| A.1 Blisters  [438] | Blisters resemble bubbles that appear at the surface of the FRP composites. | They are due to freeze–thaw cycles of trapped moisture either at the inter-laminar interfaces of the composites or at the interface between FRP composites and concrete substrate. | Since blisters are limited to the surface of FRP composites, they usually do not have a significant impact on the integrity of a structure [28, 439]. |
| A.2 Wrinkling 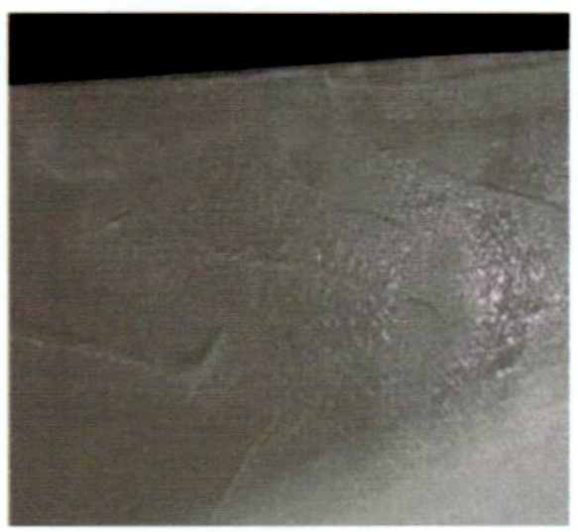 [28] | Wrinkling defects in FRP composites, having an appearance of a crease or fold, are surface defects on FRP composites. | They are caused by excessive stretching or shearing of FRP fabrics during fabrication or installation [15, 28]. | Occurring usually at corners and curves, these defects pose structural issues only if they prevent adequate bonding of the FRP composites with the concrete substrate [28, 438]. |

*(Contiued)*

**TABLE 6.1 (CONTINUED)**

**Defects/damage for external application of FRP, causes, and severity**

| Defects | Description | Causes | Severity |
|---|---|---|---|
| | | **Defects in FRP composite (Defect category A)** | |
| A.3 Scratches 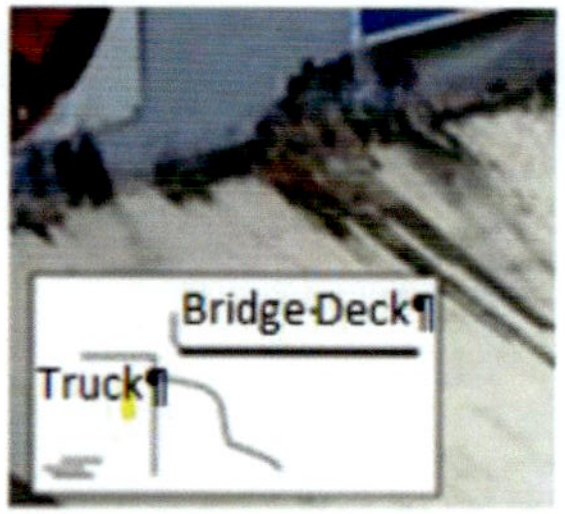  [265] | Scratches are marks or shallow cuts or even deeper grooves on the surface of FRP composites. | Scratches in externally applied FRP can occur both during installation and service due to mechanical damages. | They do not pose a significant risk to the element unless they propagate deep into the composite [176] and could be detrimental if they develop into full-depth cracks [28]. |
| A.4 Discoloration 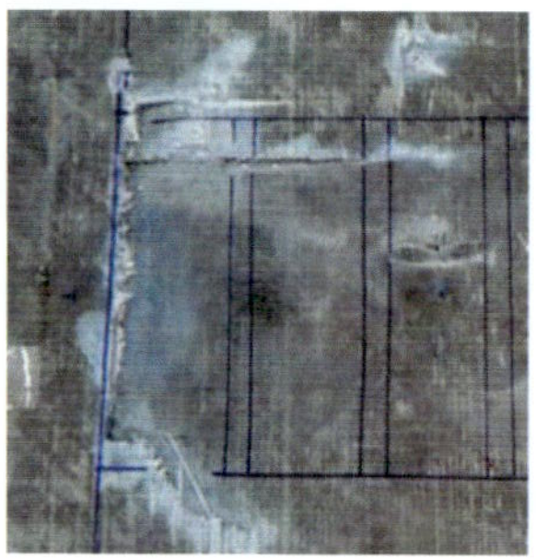 [28] | Discoloration is the change in color of the FRP composites relative to their original state or surrounding area. | Discoloration is due to weathering effects such as extended exposure to UV radiation, heat, and chemical environment [346, 438]. It can also be caused by fire, excessive strain, subsurface defects, voids, and moisture penetration [28, 438]. | UV rays cause discoloration of the matrix material through mechanisms such as photolysis and photo-oxidation, which is usually accompanied by surface cracking, embrittlement, and composite degradation [40]. |

(*Contiued*)

**TABLE 6.1 (CONTINUED)**
**Defects/damage for external application of FRP, causes, and severity**

| Defects in FRP composite (Defect category A) | | | |
|---|---|---|---|
| **Defects** | **Description** | **Causes** | **Severity** |
| A.5 Fiber exposure  [28] | Fiber exposure is a condition in which the reinforcing fibers of the FRP composites are exposed in abraded or chemically attacked areas. [440]. | Fiber exposure is due to scratches and abrasion during improper handling of the FRP composite laminates. | The exposed fiber makes the composite susceptible to ingress of moisture and contaminations and, hence, should be treated to prevent further deterioration [28]. |
| A.6 Voids at fiber–matrix interface 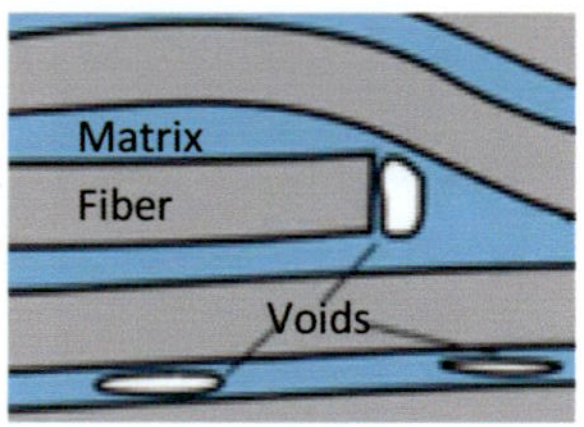  [265] | Voids at fiber–matrix interface are pockets of trapped air or gas within the FRP composites. | Voids inside FRP composites are due to entrapped air between the layers of FRP fabrics when they are laid upon each other in multiple layers. The overlapping or folding of the fabrics during fabrication of FRP laminates is another reason for their formation. | The presence of voids in laminates severely deteriorate their inter-laminar shear strength [265]. |

*(Contiued)*

## TABLE 6.1 (CONTINUED)
**Defects/damage for external application of FRP, causes, and severity**

| Defects in FRP composite (Defect category A) | | | |
|---|---|---|---|
| **Defects** | **Description** | **Causes** | **Severity** |
| A.7 Debonding at fiber–matrix interface 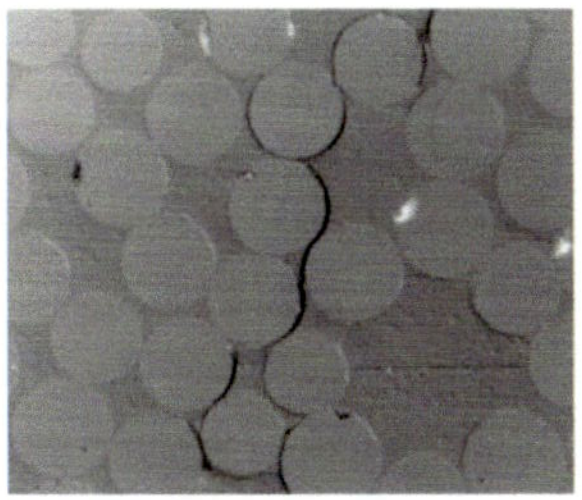 [441] | Debonding intrinsic to FRP composites is the separation at the fiber–matrix interface [15]. Debonding limited to localized areas may become larger and form delamination. | Debonding at fiber–matrix interface is due to loss of bond strength usually because of moisture present on the fiber surface. | Debonding weakens the composite action and reduces the strength of the FRP composites against transverse tension, inter-laminar shear, and impact [265]. |
| A.8 Delamination between composite layers 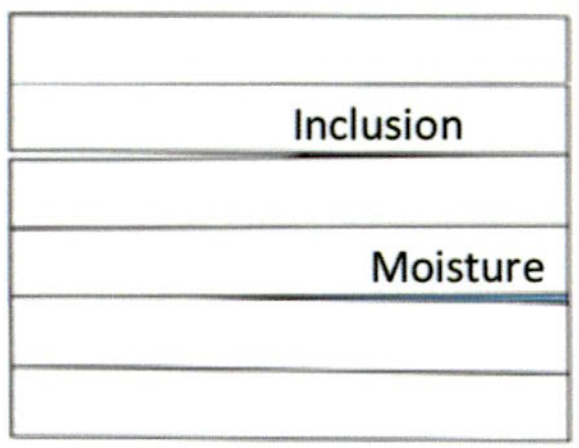  [265] | Delamination within the FRP composites is the separation of the layers of fiber-reinforced material that are stacked to form a laminate [15]. | The main reasons behind the delamination are entrapped air between the FRP layers, contamination by foreign inclusions, and moisture. Different from debonding, delamination can occur locally and over a large area. | Delamination in FRP composites causes a significant reduction in their shear transfer capability [265]. |

(*Contiued*)

**TABLE 6.1 (CONTINUED)**

**Defects/damage for external application of FRP, causes, and severity**

| Defects | Description | Causes | Severity |
|---|---|---|---|
| **Defects in FRP composite (Defect category A)** | | | |
| A.9 Cracks 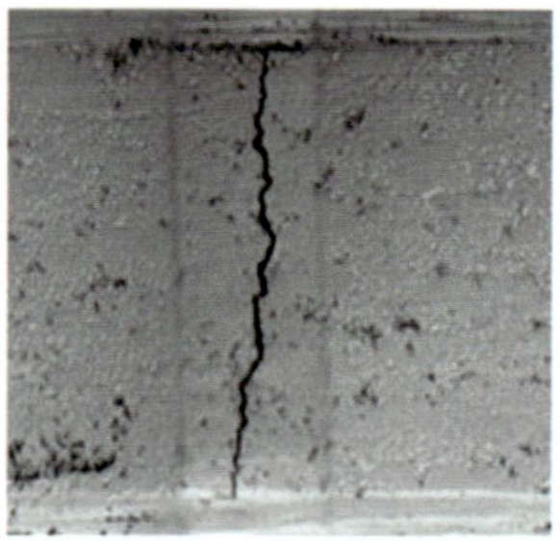 [442, 443] | Cracks in FRP composites are a fracture type of discontinuity without complete separation [440]. | Cracks are due to several internal factors, such as entrapped air, non-uniform resin distribution, and insufficient fiber reinforcement resulting in stress concentration or sudden changes in the load transfer direction [265]. In addition, FRP composites can also crack due to external factors such as impact from vehicles and debris. | Cracks may lead to the failure as they may easily propagate and grow under sustained loading, and more crucially, under load variation. |
| A.10 Impact damage 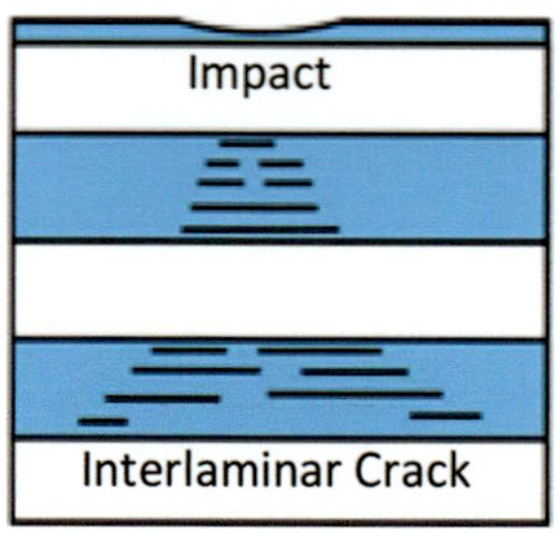  [265] | Damage caused by striking FRP composites with an object that appear as broken fibers and/or cracking are impact damage. | Impact damage can be caused due to slow-moving heavy object or vehicle (appear as minor surface damage even though the damage at subsurface level is severe), light high-velocity objects or drops (create significant surface damage). | The impact damage at the subsurface layers of the composite is more critical than surface damage as it remains hidden (barely visible impact damage). Nevertheless, surface damage may also compromise the structure's integrity. |

*(Contiued)*

**TABLE 6.1 (CONTINUED)**
**Defects/damage for external application of FRP, causes, and severity**

| Defects | Description | Causes | Severity |
|---|---|---|---|
| **Bond Defects (Defect categories B, C, and D)** | | | |
| B.1/ D.1 Debonding between FRP and concrete<br>Debonding through cement matrix or along adhesive layer<br>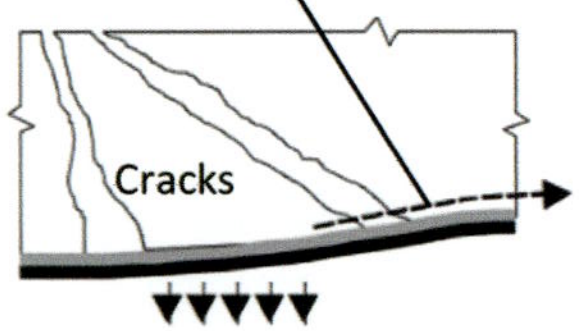<br><br>FRP pulls away from substrate<br>[5] | Debonding of externally bonded FRP composites is the failure of cohesive and adhesive bond at the interface between the FRP and the concrete substrate [444]. | Debonding of FRP composites from concrete substrate occurs as shear or flexural cracks widen up under loading and cause high interfacial shear stress [5]. Other pre-service factors such as improper installation, improper resin curing, moisture near the concrete surface also result in FRP debonding [445]. | The stress transfer from the concrete to the FRP via the bond layer is responsible for the improved performance of FRP-strengthened concrete elements. Failures in this transfer zone in the form of debonding which causes loss of force transfer may result in a brittle fracture of the concrete element [26]. |
| B.2/ C.1/ D.2 Voids between FRP and concrete<br>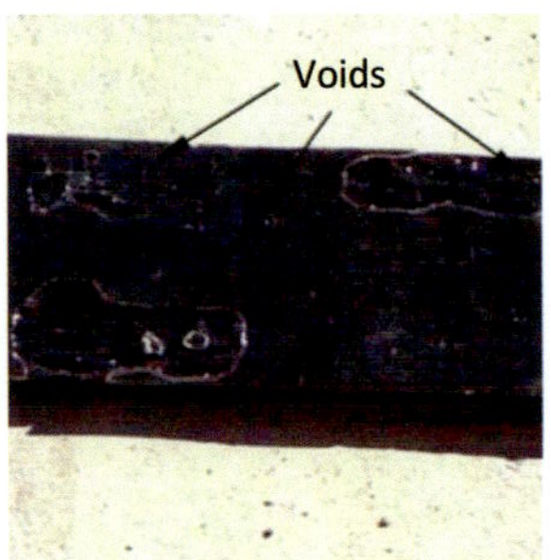<br><br>[265] | Voids between FRP and concrete are pockets of trapped air or gas at the bond layer (i.e., FRP–adhesive interface, adhesive and adhesive–concrete interface) between externally bonded FRP system and concrete. | These occur typically due to entrapped air during field installation. They are also caused due to presence of volatiles or entrapped air in resin, insufficient resin quantity, foreign particles contamination, and irregular substrate surface due to improper surface preparation [265]. | Voids are areas of zero stress transfer and cause stress concentrations in their vicinity, weakening the bond strength of the externally applied FRP system. |

*(Contiued)*

**TABLE 6.1 (CONTINUED)**

**Defects/damage for external application of FRP, causes, and severity**

| Defects | Description | Causes | Severity |
|---|---|---|---|
| **Defects in concrete (Defect categories E and F)** | | | |
| E.1 Cracks in concrete substrate 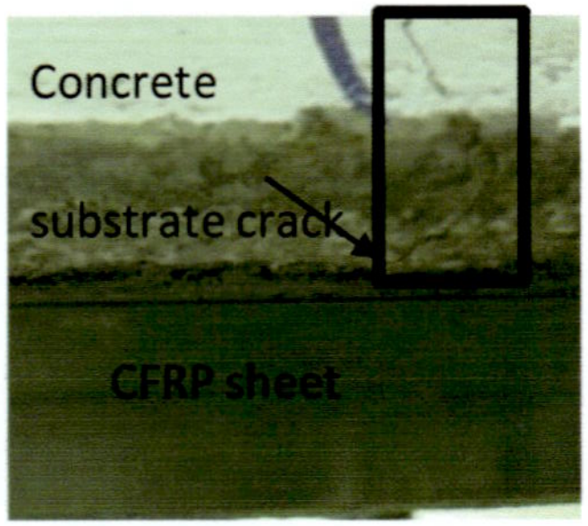  [446] | Cracks in concrete substrate are a fracture type of discontinuity without complete separation [440] and they remain hidden by externally applied FRP. | Cracks are caused due to shrinkage, thermal stresses, chemical attack, weathering, corrosion of steel reinforcement, error in design and detailing, poor construction practices, and external load [447]. | The hidden cracks that appear as hairline cracks at the beginning can grow over time and lead to failure [448]. Further, these cracks can also cause failure of the FRP strengthening system (e.g., debonding). |
| E.2 Voids in concrete 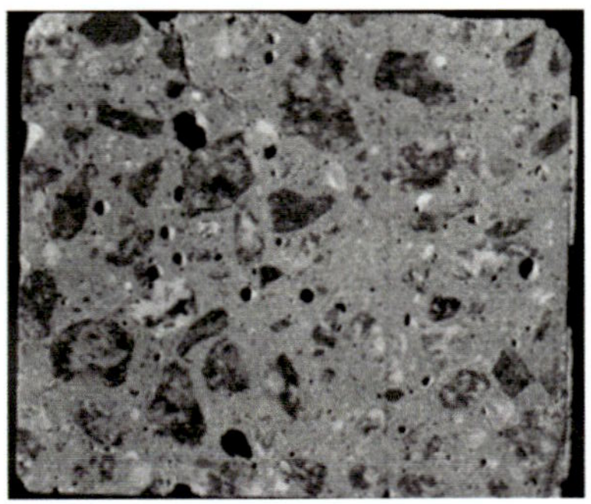 [449] | Voids are the cavities present in the concrete substrate. These could be either visible on the surface or remain hidden within the concrete element. | They form due to improper consolidation during construction [450]. Furthermore, voids can form during concrete laying due to stiff or unworkable concrete, segregation, crowded rebar, and irregular aggregate sizing, regardless of vibration quality [451]. | Voids have an adverse effect on the strength and durability of concrete [452]. They further allow easy passage to harmful chemicals that degrade the microstructure of the concrete [453]. |

**TABLE 6.1 (CONTINUED)**
**Defects/damage for external application of FRP, causes, and severity**

| Defects | Description | Causes | Severity |
|---|---|---|---|
| **Defects in concrete (Defect categories E and F)** | | | |
| E.3 Delamination 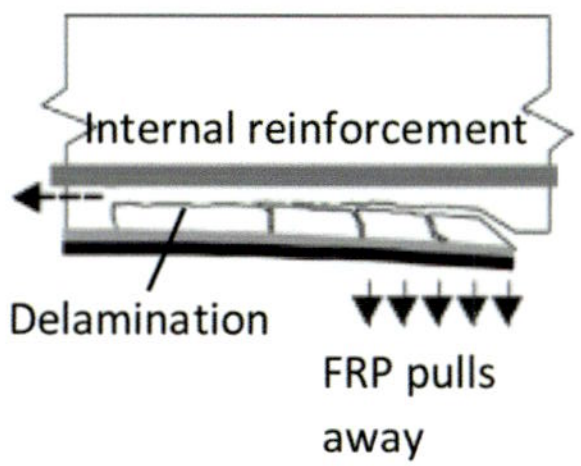  [5] | Concrete delamination of the concrete substrate (the weakest link in the FRP-adhesive-substrate region) is defined as the separation of concrete due to the effects of high stresses in the presence of externally applied FRP system. | It can occur if the substrate cannot withstand the forces in the FRP [5]. They begin with the formation of a crack at or near the end of externally bonded FRP systems due to high interfacial shear and normal stresses caused by their abrupt termination under service loads. | They are sudden, brittle, and without warning [454]. |
| F.1 Cover separation 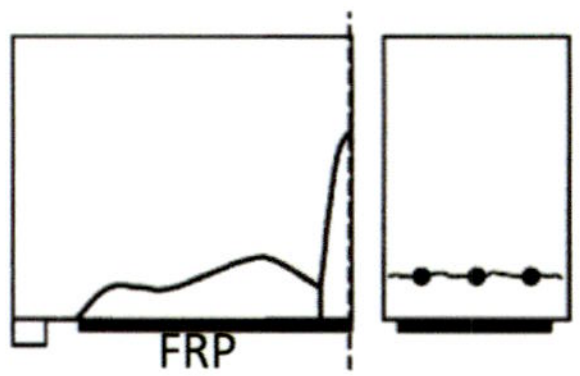  [5] | Cover separation is defined as the separation of concrete at the level of internal reinforcement due to the effects of high stresses in the presence of externally applied FRP system. | It occurs when a crack forms in the concrete at or near the termination points of FRP composites, propagates to the reinforcement level, and then progresses horizontally along the weak plane created by the presence of the reinforcement. | They are sudden, brittle, and without warning [454]. |

*(Contiued)*

**TABLE 6.1 (CONTINUED)**
**Defects/damage for external application of FRP, causes, and severity**

**Defects in concrete (Defect categories E and F)**

| Defects | Description | Causes | Severity |
|---|---|---|---|
| F.2 Corrosion in steel reinforcement  [455] | Corrosion of steel reinforcement in concrete is defined as its degradation as a result of electrochemical reaction with its environment. | It is caused due to reduction of pH of the concrete due to carbonation or other mechanisms, chloride contamination, and diffusion of halides and chemicals corrosive to reinforcing steel. Although the use of external FRP may reduce the rate of corrosion in the embedded steel reinforcement, corrosion cannot be avoided [456]. | Corrosion in FRP-strengthened concrete has the same adverse effects as in conventional reinforced concrete. Hence, assessing post-repair corrosion is critical in order to justify the use of FRP in repairing corrosion-damaged elements [457] |

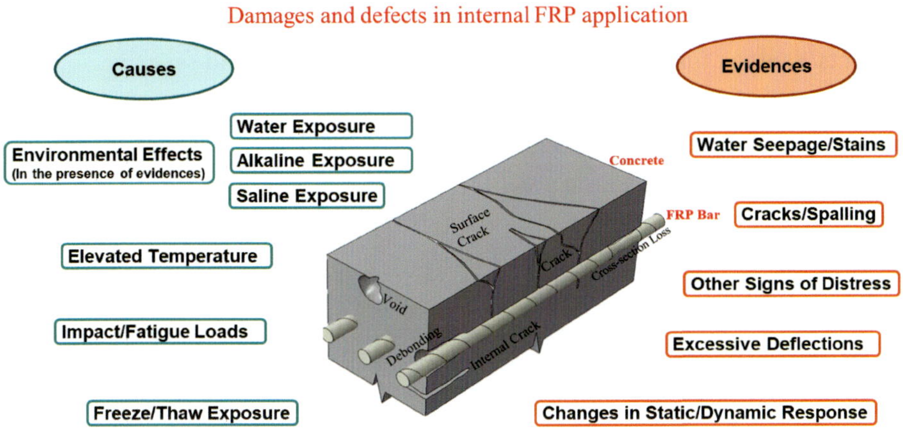


**FIGURE 6.2** Causes and evidence for potential deficiencies in internal FRP application

needed to ensure construction quality and continued structural integrity. NDT is vital because the failure of structures reinforced with FRP rebars is not as ductile as conventional constructions, and concrete structures strengthened with externally bonded FRP composites are covered and cannot be inspected as readily. NDT methods that can inspect beyond the FRP wraps or laminates to examine the internal integrity of the element should be evaluated.

### 6.3.1 Inspection Techniques

This section reviews inspection and non-destructive testing techniques that could potentially be applicable to FRP inspection summarized based on a literature survey of past studies, applications, and research projects. These methods were selected from a large pool of NDT methods available for damage detection in structural elements. Table 6.3 provides a brief overview of selected methods in relation to FRP-RSC, including their descriptions, applications, advantages, and disadvantages. This information serves as the foundation for choosing the most suitable methods. For each NDT technique, its theoretical principles and applicability are discussed in detail in the previous chapters.

### 6.3.2 Selection of Applicable NDT Methods

The following flowchart (Figure 6.3) has been developed as the result of comparing NDT methods and their applicability to FRP-RSC elements. It shows NDT methods that are potentially applicable to FRP-RSC and suggests the order of priority of various NDT methods, types of deficiencies, and types of applications.

**TABLE 6.2**
**Defects/damage for internal application of FRP, causes, and severity**

**Defects in FRP composites (Defect categories A and G)**

| Defects | Description | Causes | Severity |
|---|---|---|---|
| G.1 Loss of cross-sectional property | Loss of cross-sectional property is the deterioration of internal FRP reinforcement that results from chemical reaction with its environment. | It is caused mainly due to chemical attacks from alkaline, acidic, saline, and other corrosive environments degrading the constituents (viz., fibers and matrix), as well as the interface in FRP composites (viz., fiber–matrix interface). | It reduces the service life of FRP-reinforced concrete member. However, there might not be any visible or physical signs of chemical attacks despite the reduction in their mechanical properties [458]. |
| A.1–A.9 (Same as Fig. 2a) | Defects **A.1** Voids at fiber–matrix interface, **A.2** Wrinkling, **A.3** Blisters, **A.4** Debonding at fiber–matrix interface, **A.5** Delamination between composite layers, **A.6** Fiber exposure, **A.7** Scratches, **A.8** Cracks, and **A.9** Discoloration have been already discussed in Fig. 2a and can also occur in the internal FRP reinforcement during manufacturing, transportation, storage and handling, and service. | | |

**Defects at the interface (Defect category H)**

| Defects | Description | Causes | Severity |
|---|---|---|---|
| H.1 Debonding | Debonding is the deterioration of the interfacial bond between the internal FRP reinforcement and concrete [459]. | Due to environmental and load factors, the bond may gradually weaken over time, resulting in bond failure of the FRP-reinforced concrete parts, and jeopardizing the structure's integrity [276], [459, 460]. | The bond between concrete and FRP reinforcement ensures that the high tensile strength of FRP rebars is used effectively [461]. Debonding might compromise structure's integrity and tensile resistance making it susceptible to more damage [459]. |

*(Contiued)*

**TABLE 6.2 (CONTINUED)**
**Defects/damage for internal application of FRP, causes, and severity**

| Defects | Description | Causes | Severity |
|---|---|---|---|
| **Defects at the interface (Defect category H)** | | | |
| H.2 Slippage | Slippage is the relative displacement of the internal FRP reinforcement when external forces overcome bond with concrete [462]. | It occurs when external forces overcome bond strength between the FRP reinforcement and concrete [462] which is usually lower than that of conventional steel rebars [463]. | Bond-slip is critical as it weakens the structural integrity and overall strength of FRP-reinforce concrete elements. Thus, detection and prediction of bond-slip events are important to issue early warnings of possible failure [464]. |
| **Defects in concrete (Defect category I)** | | | |
| **Defects** | **Description** | **Causes** | **Severity** |
| I.1 Cracks in FRP-reinforced concrete  [465] | Cracks are a fracture type of discontinuity without complete separation. In FRP-reinforced concrete they are mostly limited for aesthetics of the structure, unlike steel reinforced concrete, in which corrosion is the main concern [178, 466]. | Cracks in FRP-reinforced concrete elements can be attributed to the lower modulus of elasticity of FRP rebars [465] as compared to conventional steel. | Extensive cracks and large deflections generally occur prior to failure in the concrete members reinforced with FRP rebars [466] which give the warning of impending failure [467]. |
| I.2—I.3 (Same as Fig. 2c) | The defects in concrete such as **E.2** Voids and **E.3** Delamination as described in Fig. 2c are the same whether FRP is used for strengthening or reinforcing purposes. | | |

## 6.4 INSPECTION TYPE

The type of inspection to be applied depends upon the condition of the bridge. It is important to understand the inspection types for adequate inspection of any bridge. The types of inspection as defined by NBIS 23 CFR § 650.305 are briefly introduced in this section.

### 6.4.1 Initial Inspection

The first inspection of a new, replaced, or rehabilitated bridge. This inspection serves to record bridge inventory data, establish baseline conditions, and establish the intervals for other inspection types.

**TABLE 6.3**
**NDT methods applicable for FRP inspection [468]**

| Method | Description | Advantages | Disadvantages |
|---|---|---|---|
| Visual Inspection (VT)<br> | • The most common, versatile, and simplistic NDT method that can be used for inspecting FRP-RSC elements,<br>• Defects such as cracks, delamination, fractures, blisters, debonding, moisture accumulation, fiber misalignment, discoloration or any other surface defects can be detected by visual inspections. | • Checks obvious and outstanding anomalies,<br>• Fast, simple and cost effective,<br>• Real-time inspection and can be interpreted instantly,<br>• Sets the baseline for other NDT methods,<br>• Based on the findings of visual inspection, decisions can be made on necessity of further inspection with more sophisticated NDT methods,<br>• Can be performed before any other NDT as a supplementary method. | • Detects only surface defects,<br>• Subjective and observations may vary depending on visual perceptions of individuals. |
| Tap Testing (TT)<br> | • The local stiffness of a laminated material within a defective area differs from that in sound areas which causes a variation in the frequency of excitation upon impact,<br>• In this method, defects (i.e., voids, debonding, delamination) are determined by tapping the test object with a coin or a hammer, and then listening for a change in (frequency of) the sound produced. | • Faster inspection,<br>• Can inspect large areas,<br>• Low cost,<br>• Easy to use method with some training,<br>• Real time and provides immediate results. | • Subjective as it relies on the human factors,<br>• Results vary due to the inconsistencies in the force, angle and devices used,<br>• Ambient noise and geometric changes may lead to erroneous interpretation,<br>• Does not provide the depth and width of the defects. |

(*Contiued*)

**TABLE 6.3 (CONTINUED)**
**NDT methods applicable for FRP inspection [468]**

| Method | Description | Advantages | Disadvantages |
|---|---|---|---|
| Impact Echo (IE)<br>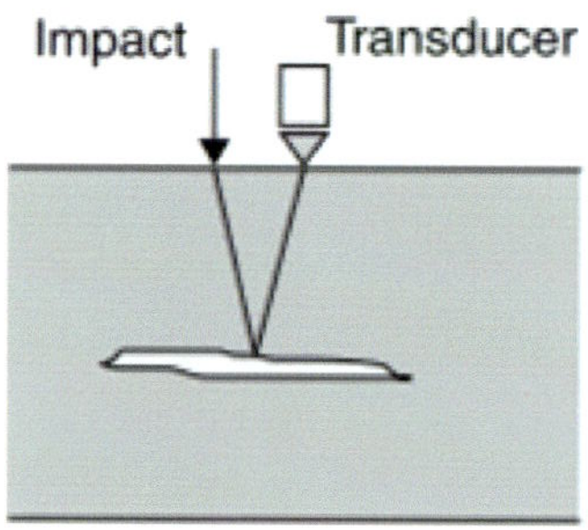<br> | • Uses stress waves or mechanical waves generated by an impact (by a small hammer or a steel sphere) that propagate through the member and reflect back due to variation in acoustic impedances caused by the presence of discontinuities,<br>• The reflected waves recorded by a displacement transducer are analyzed to give information about the extent and location of the subsurface defects. | • Has deep penetrating capability into concrete,<br>• Promising for evaluation of cracks, voids, delamination, and discontinuities in concrete,<br>• Unlike ultrasonic methods, impact echo utilizes lower frequency which allows the impact echo to overcome high-signal attenuation,<br>• Needs only one surface to be accessible for testing. | • Current instrument limited to testing members up to 40 in. (1 m) thick,<br>• Operate by experienced personnel,<br>• Smaller cracks and discontinuities cannot be detected easily. |
| Ground-Penetrating Radar (GPR)<br>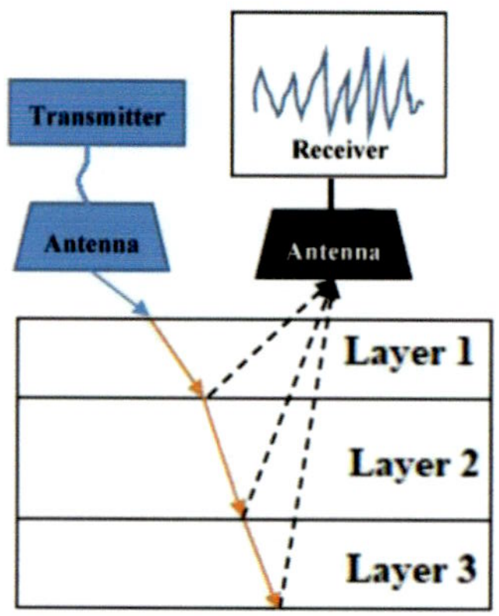<br> | • Operates by transmitting electromagnetic waves (radio waves) through the test material and by receiving the waves that are reflected off the discontinuities within the material,<br>• These discontinuities could be any interface between different materials (such as concrete–rebar interface) or subsurface defects like voids, cracks, debonding, and delamination (concrete–air interface or concrete–water interface). | • Penetrates beyond concrete–air interfaces which allows inspection of features below the interface,<br>• Detects defects at a greater depth than some other NDT techniques such as infrared thermography,<br>• Air-coupled GPR allow rapid inspection. | • Relatively more expensive than other NDT methods,<br>• Not suitable for detecting air-filled defects. |

*(Contiued)*

## TABLE 6.3 (CONTINUED)
### NDT methods applicable for FRP inspection [468]

| Method | Description | Advantages | Disadvantages |
|---|---|---|---|
| Radiographic Testing (RT)<br>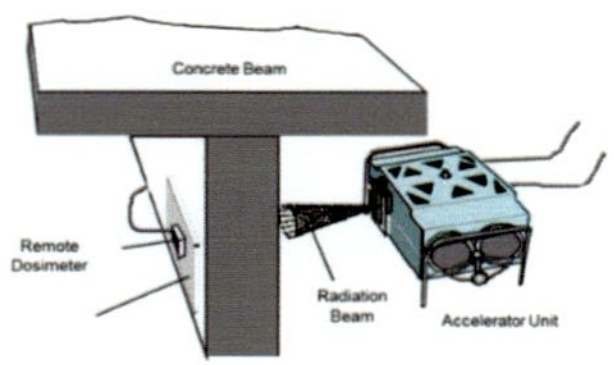<br> | • Operates by recording the intensity of high-energy electromagnetic radiation, such as X-rays and Gamma rays, on a photographic film after passing them through the test material,<br>• The differences in the intensity of the radiation that can be seen as shadows or bright spots in the photographic film give information about the internal features of the member such as voids or delamination. | • Can clearly detect the defects like delamination in the FRP as there would be variations in the absorption of the radiation between the sound and defective regions,<br>• Defects that cause significant density variations, such as excess density of fiber or matrix, resin variations, impact damage, delamination, debonding, foreign contamination, cracks, and voids, can be easily detected,<br>• Can also detect non-uniform fiber distribution, broken fiber, poor fiber weaving, and misorientation in fibers, such as fiber wrinkles. | • If the orientation of the delamination is perpendicular to the radiations, these defects will not be detected in a 2D scan,<br>• More expensive than other NDT methods,<br>• Safety issues from high intensity radiations. |

(Contiued)

**TABLE 6.3 (CONTINUED)**

**NDT methods applicable for FRP inspection [468]**

| Method | Description | Advantages | Disadvantages |
|---|---|---|---|
| Ultrasonic Testing (UT) 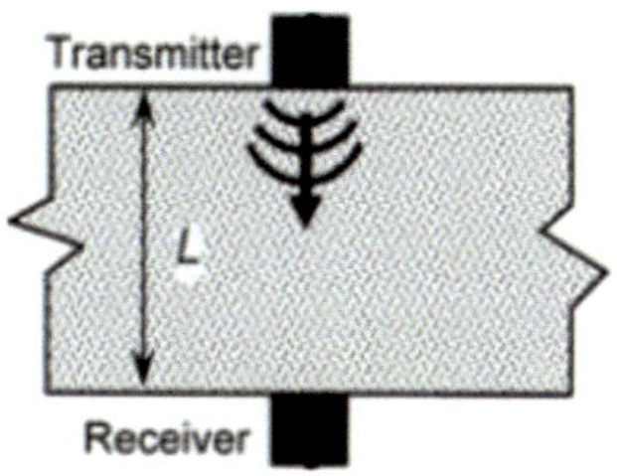  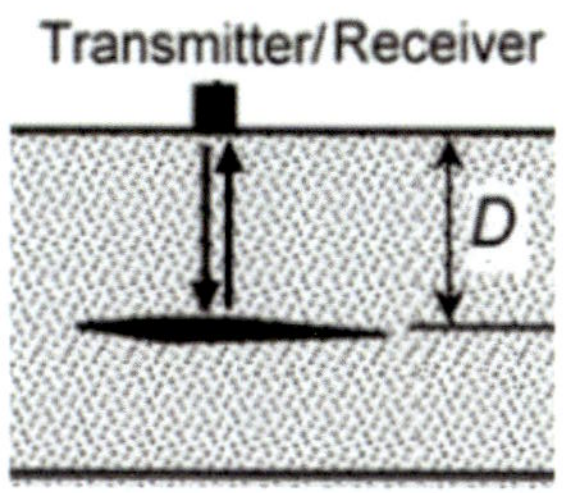  | • Works based on the principle that when the ultrasonic waves come across an interface between two materials with different acoustic impedances, a portion of the incident wave will be reflected back,<br>• Has been very successful in locating defects within concrete and composites as there is almost a total reflection at the interface present due to defects. | • Fast, easy to use in fields, and has good resolution,<br>• Can penetrate well into the materials and detect subsurface flaws,<br>• Detects wide range of defects such as debonding, resin variations, broken fibers, impact damage, moisture, cracks, voids, and subsurface defects. | • Highly trained individual to conduct the test and interpret the data,<br>• Applicable to limited thickness of members. |

*(Contiued)*

**TABLE 6.3 (CONTINUED)**
**NDT methods applicable for FRP inspection [468]**

| Method | Description | Advantages | Disadvantages |
|---|---|---|---|
| Infrared Thermography (IR) 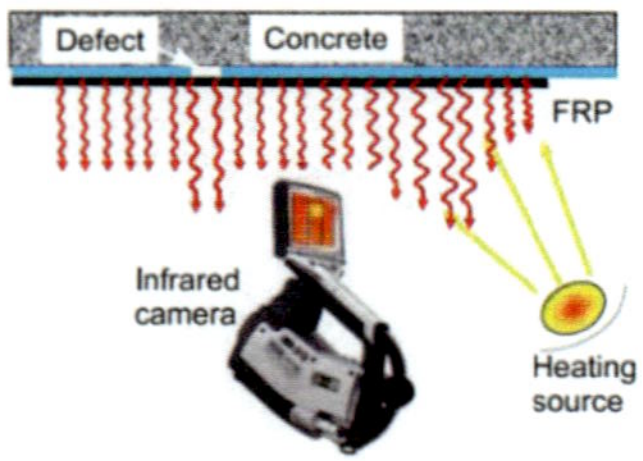  | • Works based on the principle that the thermal properties (thermal conductivity) of the anomalies present within the element are different than that of the surrounding sound part,<br>• It can detect subsurface anomalies to some extent by measuring the surface temperature. | • Detects subsurface defects globally over larger surface areas,<br>• Fast and inexpensive,<br>• Inspections and data interpretation can be performed in real time. | • Not reliable enough in detecting water-filled defects,<br>• Cannot effectively detect defects present deep within the concrete,<br>• Needs proper environmental/ambient conditions. |
| Acoustic Emission (AE) 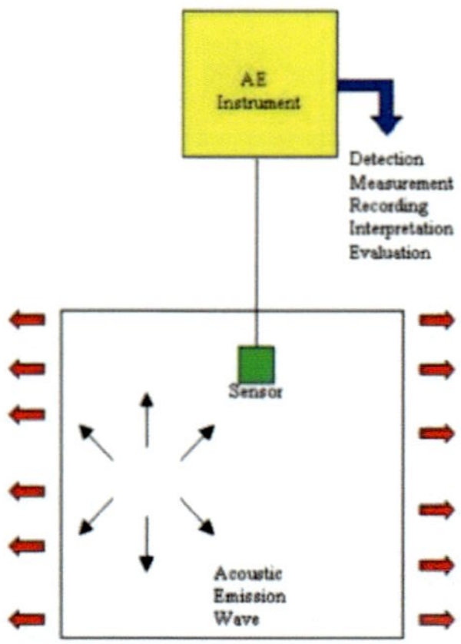  | • Operates by detecting the elastic waves generated by sudden redistribution of stress produced due to moving dislocations, onset/growth/propagation of cracks, fiber breaks, debonding, and plastic deformations. | • Allows fast and global inspections,<br>• Can be performed without interruption to the service of the structure,<br>• Detects defects due to fatigue loading such as fatigue cracks, fiber fractures, matrix micro-cracks, fiber-matrix debonding, and delamination,<br>• Effective in locating the origin/location of discontinuities,<br>• Detects the growth rate of a defect. | • Difficulty in differentiating deficiency source,<br>• Needs high level of skill in correlating the AE data to respective deficiency sources,<br>• Noisy environment may lead to false signals,<br>• Defects might go undetected if the loading is not sufficient,<br>• Cannot give information on the type and size of defects,<br>• Not capable of detecting deficiencies occurred before sensor installation. |

(Contiued)

**TABLE 6.3 (CONTINUED)**
**NDT methods applicable for FRP inspection [468]**

| Method | Description | Advantages | Disadvantages |
|---|---|---|---|
| Global Structural Response Testing (GSR)<br>• Modal Testing<br>• Load Testing<br> | • The principle behind modal testing is to evaluate the condition of a structure by monitoring the changes in its dynamic response/ behavior (typically, modal frequencies and modal shapes),<br>• The principle behind load testing is to evaluate the condition of a structure by monitoring the changes in its static response such as increase in deflection of a structure under externally applied load. | • Effective for assessing the efficacy of the FRP rehabilitations,<br>• Applicable for long-term performance of structures. | • The local defects may not be significant enough to contribute to global response,<br>• Lacks sensitivity to finer localized defects and cannot give information on their size, location, and magnitude. |
| Complementary to NDT Method: Vision-based Techniques<br>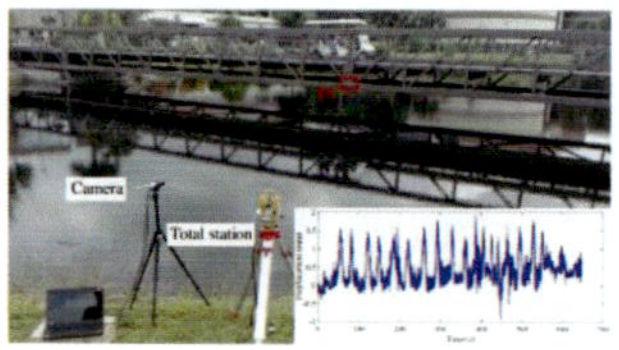 | • Uses photographs and videos as the primary means of analysis and assessment,<br>• At the local level, it can detect surface defects such as cracks and delamination, as well as internal structural issues through the use of visual and infrared imaging,<br>• At the global level, it can be used to analyze the overall structural health by tracking and analyzing parameters such as displacement, vibration, and modal behavior. | • Non-contact measurements,<br>• Remote data collection,<br>• Low cost and less time-consuming,<br>• Less labor-intensive,<br>• Minimum interference to traffic. | • Precision can be affected by vibrations and movement of camera,<br>• The field of view could be distorted due to environmental factors which can give erroneous observations. |

(*Contiued*)

**TABLE 6.3 (CONTINUED)**

**NDT methods applicable for FRP inspection [468]**

| Method | Description | Advantages | Disadvantages |
|---|---|---|---|
| Complimentary to NDT Method: Robotic Inspection<br> | • Involves the use of NDT techniques integrated with robots,<br>• The robots used for NDT application can be divided into aerial robots, ground robots, and underwater robots. | • Less time-consuming,<br>• Automated robotic inspections are quite efficient,<br>• Can reach places where humans cannot,<br>• Reduces the risk of accidents involving humans. | • Environmental factors such as improper lighting, ambient noise, high wind speed, weather conditions, low visibility, etc. affect vision-based robotic inspection,<br>• Restrictive regulations in flying UAVs,<br>• Difficulty in maneuvering robots in congested space. |

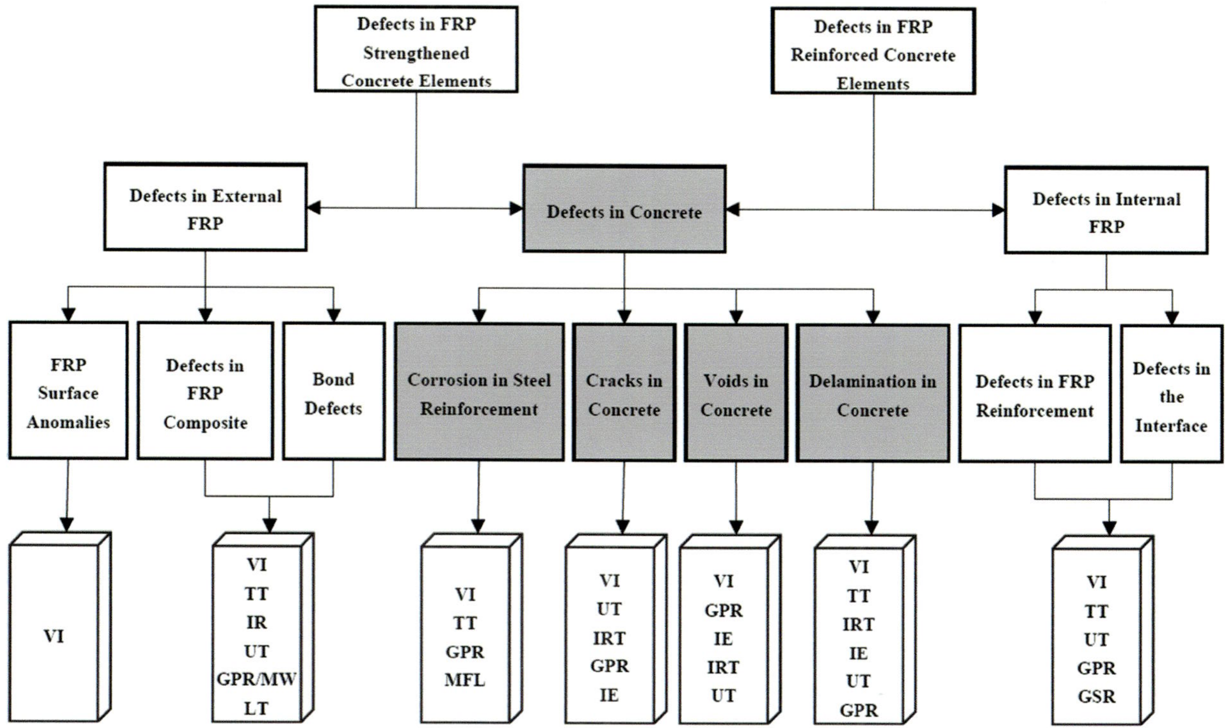


**FIGURE 6.3** Order of priority for NDT methods suitable for each type of defect

Notes: 1. Defects within FRP composite layers include blisters, wrinkling, scratches, discoloration, fiber exposure, voids at the fiber–matrix interface, fiber–matrix debonding, delamination between composite layers, cracks, and impact damage

2. Bond defects include debonding and voids between FRP and concrete

3. Corrosion in steel reinforcement is to be considered only when it is applicableAcronyms:VT = Visual Testing, TT = Tap Testing, IE = Impact Echo, MW = Microwave, GPR = Ground Penetrating Radar, UT = Ultrasonic Testing, LT = Laser Testing, IR = Infrared Thermography Testing, IRT = Impulse Response Testing, MFL = Magnetic Flux Leakage, GSR = Global Structural Response Testing

### 6.4.2 Routine Inspection

Regularly scheduled comprehensive inspection consisting of observations and measurements needed to determine the physical and functional condition of the bridge and identify changes from previously recorded conditions. The routine inspection interval should follow NBIS.

### 6.4.3 Damage Inspection

An unscheduled inspection to assess structural damage resulting from environmental factors or human actions.

### 6.4.4 In-Depth Inspection

A close-up, detailed inspection of one or more bridge members located above or below water, using visual or nondestructive evaluation techniques to identify any deficiencies not readily detectable using routine inspection procedures. Hands-on inspection may be necessary at some locations.

### 6.4.5 Special Inspection

An inspection scheduled at the discretion of the bridge owner, used to monitor a particular known or suspected deficiency, or to monitor special details or unusual characteristics of a bridge that does not necessarily have defects.

### 6.4.6 Other Inspection

Hands-on inspection: Inspection within arm's length of the member. It uses visual techniques that may be supplemented by nondestructive evaluation techniques.

Underwater inspection: Inspection of the underwater portion of a bridge substructure and the surrounding channel, which cannot be inspected visually at low water or by wading or probing, and generally requiring diving or other appropriate techniques.

## 6.5 INSPECTOR QUALIFICATIONS, SITE SAFETY, AND OTHER CONSIDERATIONS

To ensure proper inspection of structures, inspection team leaders must meet the qualifications outlined in the Code of Federal Regulations (23 CFR 650.309), the most recent edition of the AASHTO Manual for Bridge Evaluation, and as specified by the state DOTs. Furthermore, inspectors should have experience and be familiar with FRP material and able to recognize the need for advanced inspection methods. The inspectors should be able to analyze and interpret the resulting data. Nondestructive testing (NDT) engineers employed by the highway organizations or consulting firms may conduct some of the highly specialized inspections.

To ensure safety and compliance during inspections, the team leader should plan and oversee traffic control, access, equipment, and other site-related matters. This should be performed in accordance with regulations and guidelines, including those of the U.S. Department of Labor's Occupational Safety and Health Administration, the Manual of Uniform Traffic Control Devices, and the AASHTO Manual for Bridge Evaluation. Additionally, state or local regulatory authorities may have their own specifications, which should be observed when addressing issues associated with fieldwork.

## 6.6 INSPECTION PROCEDURE

Tables 6.1 and 6.2 classified deficiencies in two main categories of FRP application. Accordingly, the inspection of FRP-RSC elements is discussed separately for the same two categories, FRP-strengthened bridge elements (external application) and FRP-reinforced bridge elements (internal application).

### 6.6.1 FRP-Strengthened Bridge Elements (External Application)

The inspection of FRP-strengthened bridge elements can be conducted for three different categories of deficiencies. The first category refers to the deficiencies that are readily visible on the surface and fall under the FRP composite defect mentioned in Section 4.1.1 of Chapter 4. The second category refers to the deficiencies that occur in the FRP composite itself and at the bond layer between the FRP and the concrete. Finally, the third category refers to the deficiencies in the concrete substrate. These were shown in Figure 6.1 and Table 6.1.

#### 6.6.1.1 Inspection for Surface Anomalies and Defects

The externally applied FRP system should be thoroughly investigated visually. Although the focus of the visual inspection is the externally bonded FRP, the overall condition of the bridge should also be visually assessed. As a minimum, the following should be checked during the inspection:

- Look for signs of surface anomalies such as blisters, fiber exposure, scratches, and cracks on the surface of the externally applied FRP (Figures 6.4, 6.5, and 6.6).
- Look for signs of moisture, water seepage, efflorescence, and water stains (Figures 6.7 and 6.8) usually appearing near joints or lower areas underneath the structures.
- Check for regions of discoloration, usually occurring at areas exposed to UV rays but could be from other exposures, even from impact (Figure 6.9).
- The surface anomalies observed in externally applied FRP could be an indication of defects within the FRP composite or the bond defects between the FRP and concrete. Use suggested NDT methods mentioned in Section 6.3.2 for determining the extent of the observed deficiencies if warranted.

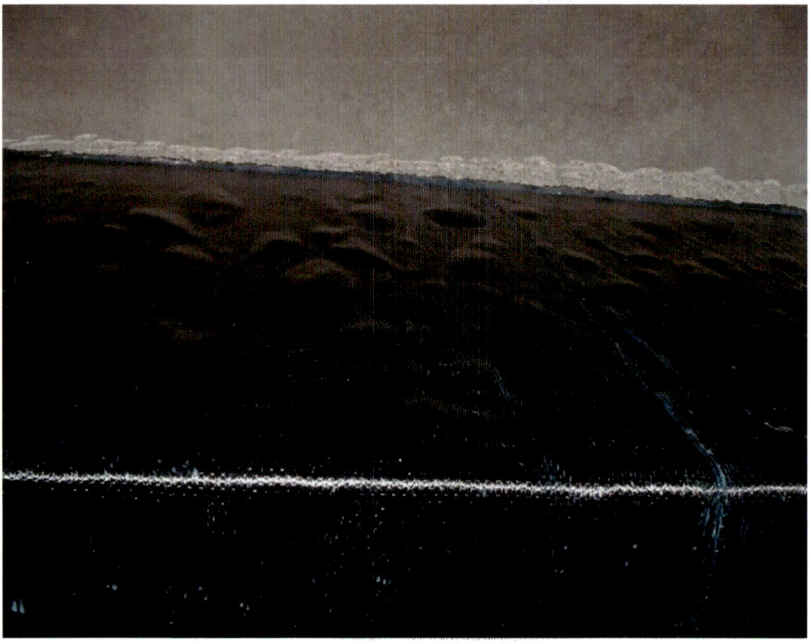

FIGURE 6.4 Blistering/Bubbles in FRP

FIGURE 6.5 Fiber exposure [469]

**FIGURE 6.6** Cracks

#### 6.6.1.2 Inspection for Defects in FRP Composite and Bond Issues

The inspection for defects in FRP composite and bond defects may be evaluated by NDT methods other than visual inspections. Tap testing should be used for bond damage detection. If warranted, IR, GPR, or UT can be used in addition for quantitative assessment of the deficiencies present within FRP composite or in between FRP and the concrete substrate. The following steps should be performed as a minimum:

- Check for signs of debonding and delamination in externally applied FRP. Look for evidence of loosely bonded FRP that could be effortlessly peeled from the concrete substrate. Attempt to determine whether the defect is delamination within the layers of FRP composite or debonding of the FRP from the concrete substrate. Use tap testing to identify the location of debonding/delamination in addition to visual inspection. Figures 6.10 and 6.11 show examples of debonding and end peeling.
- Look for potential signs of voids. The voids could be either visible at the surface, as shown in Figure 6.12, or show some indication of their presence,

**FIGURE 6.7** Rupture due to moisture entrapment

**FIGURE 6.8** Efflorescence near joints of adjacent box beam girders [469]

**FIGURE 6.9** FRP discoloration [470]

**FIGURE 6.10** Debonding [469], [470]

FIGURE 6.11 End peeling [469]

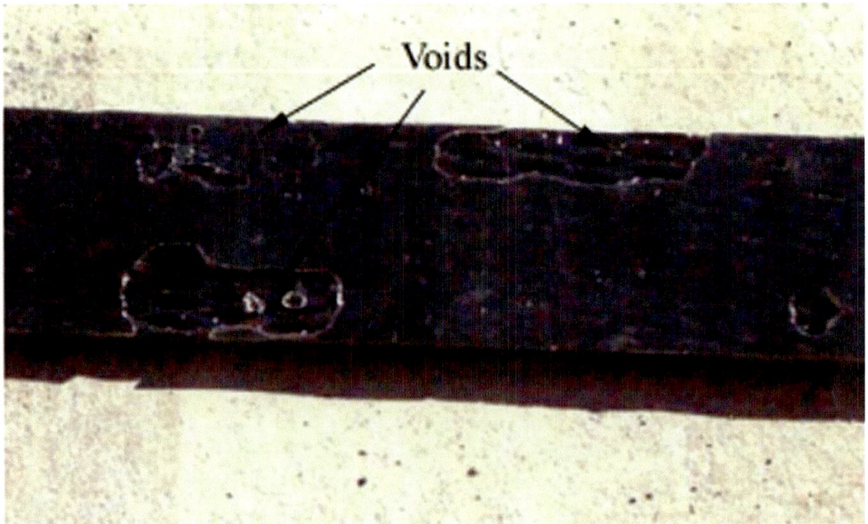


FIGURE 6.12 Void [265]

**FIGURE 6.13** Potential void [469]

as shown in Figure 6.13. The areas suspected of having void (Figure 6.13) should be further investigated using tap testing or IR.

- If warranted, detailed investigation of debonding, delamination, and voids can be conducted using NDT methods suggested in Section 6.3.2.

#### 6.6.1.3 Inspection for Defects in Concrete Substrate

The inspection of concrete hidden beneath the externally applied FRP composite is a challenging task. However, this can be achieved through inspection of evidence that indicates the presence of defects in the underlying concrete. The inspection of concrete substrate, at a minimum, should include the following activities:

- Look for signs of FRP tearing due to spalling of the concrete substrate, as shown in Figure 6.14.
- Look for the signs of abnormalities, mostly deviations from the surrounding sound external FRP regions, that would indicate the presence of a defect underneath the FRP. Figure 6.15 shows bulging of the CFRP layer along the soffit of a girder, indicating the presence of a longitudinal crack at the substrate.
- Check for signs of rust stains, as shown in Figure 6.16. Rust stains over externally applied FRP indicate the corrosion of steel reinforcement embedded in the FRP-strengthened concrete elements.
- NDT devices that can penetrate through the external FRP composite layer should be used at suspected areas for further investigation.

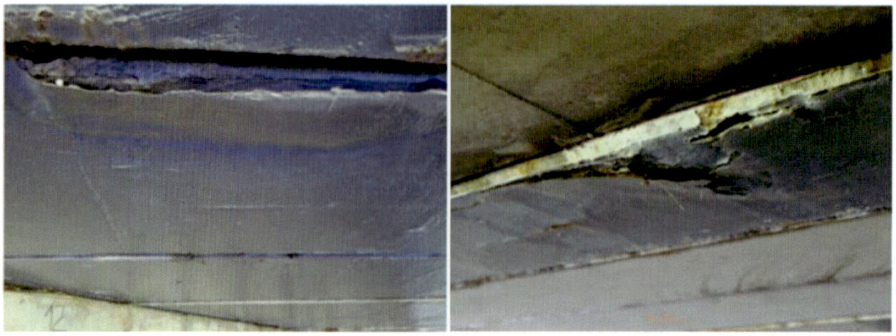

**FIGURE 6.14** Spalling of substrate concrete causing tearing of externally applied CFRP [469]

**FIGURE 6.15** Bulged CFRP layer indicating the presence of a longitudinal crack beneath [469]

### 6.6.2 FRP-Reinforced Concrete Bridge Elements (Internal Application)

In general, the inspection of reinforced concrete (RC) elements regardless of the type of reinforcement has been covered widely by the bridge inspection and evaluation manuals and guidelines. The inspection of FRP internal application differs from the conventional steel-RC, as many of the serviceability issues related to conventional RC such as cracking, permeability, carbonation, chloride content, and concrete cover may not pose the same concern for FRP-reinforced elements. On the other hand, FRP materials may be prone to such problems as creep rupture and alkaline exposure that

**FIGURE 6.16** Rust stains indicating corrosion of steel reinforcement in concrete substrate [267]

are of no concern for steel rebars. The following offers guidance for the inspection of FRP-RC elements beyond or complementary to the procedure used for conventional steel-RC elements. Inspection of FRP-PC (FRP prestressed concrete) has not been covered in this framework.

Because internal FRP is hidden from eyes, the inspection of FRP-reinforced bridge elements involves looking for the cause (sources mentioned in Section 4.4) and the respective evidence or indications of the deficiencies in the element. Since the deficiencies in FRP-reinforced concrete elements are not as obvious as FRP-strengthened concrete elements, their presence can be indirectly inferred from the signs of abnormalities (such as water seepage/stains, cracks, spalling, deflections, etc.), especially if occurring together with the potential causes/sources of deficiencies (such as excessive water/alkaline exposure, elevated temperature, extreme loading conditions, etc.), as shown in Figure 6.2. For instance, with reference to Figure 6.2, instead of directly detecting the debonding of FRP bar (shown in the center image), the inspector can look for relevant evidence such as excessive deflection (shown under the evidence column), the unusual causal factors such as extreme loading conditions (shown under the causes column), or both. The potential deficiencies in FRP before completion of the bridge that might have occurred during manufacturing/installation, transportation, storage, and handling of FRP material are not considered or covered in this framework unless they become evident due to the presence of other in-service deficiencies. Further, when and if a relationship between the cause/evidence and potential deficiencies is established, NDT inspection on suspected areas can be conducted using the methods suggested in Section 6.3.2 if warranted.

#### 6.6.2.1 Inspection of Concrete

For the inspection of concrete, as stated above, general purpose bridge inspection manuals and references along with well-established procedures for conventional reinforced concrete elements should be followed. Prior chapters in this book also described the inspection and NDT methods that are most applicable to concrete and concrete elements and can be referenced for the purpose of this framework.

#### 6.6.2.2 Inspection of Internal FRP

As mentioned earlier, the inspection of internal FRP can be carried out by attention to the causes and evidence. It should be noted that in most cases, the presence of detrimental environmental effects would not necessarily indicate a potential for deficiency in FRP if evidence is not also present. On the other hand, in some cases, the presence of evidence alone could be suspected as internal FRP deterioration such as the case of cracks and excessive deflections. Some of the evidence to look out for with their respective causes are described as follows:

- Look for signs of water seepage into the element, as shown in Figure 6.17, as it could mean that the water has penetrated to the level of the internal FRP reinforcement creating potential for deterioration in their mechanical

**FIGURE 6.17** Water seepage [471]

properties. If supported by the presence of stains, this could indicate saline exposure. However, moisture exposure is not a cause of deterioration.

- Check for cracks in the FRP-reinforced concrete elements, especially on bridge decks. Cracks are the most common types of damage occurring in the FRP-reinforced concrete elements due to the lower modulus of elasticity of the FRP bars. The cracks visible on the surface of the bridge decks reinforced with FRP rebars are shown in Figure 6.18.
- Check for any signs of distress such as fire damage with excessive spalling or burn marks, as shown in Figure 6.19.
- Deflection measurement as shown in Figure 6.20 under live load or dead load alone can be carried out to assess the overall performance of the FRP-reinforced bridges. FRP-reinforced concrete elements exhibit higher deflections than steel-reinforced concrete elements of the same size, shape, and reinforcement ratio. Having deflections higher than the design deflection limits specified by AASHTO can be construed as a potential presence of damage in the internal FRP.
- Deficiencies in FRP-reinforced concrete elements can also be detected by using damage detection methods based on variations of bridge dynamic response (natural frequency and mode shapes), as shown in Figure 6.21.
- The presence of damage and its location, type, and severity in the internal application of FRP can be further verified by using NDT methods suggested in Section 6.3.2.

## 6.7 RECORDKEEPING

### 6.7.1 Procedure for Recording Observations

To ensure that any deficiency of the FRP-RSC elements is well-documented, it is essential to record observations during the inspection that clearly indicate the location, extent, type, and severity of the deficiency. A consistent and standardized method of recording observations provides a means to compare observations using uniform evaluation criteria, which facilitates the assessment of the severity of observed conditions and the determination of appropriate action. Sufficient details and descriptions should be included in the notes to be meaningful and useful for

FIGURE 6.18 Cracks in FRP-reinforced bridge decks [465]

FIGURE 6.19 Fire damage [472]

FIGURE 6.20 Deflection measurement [473]

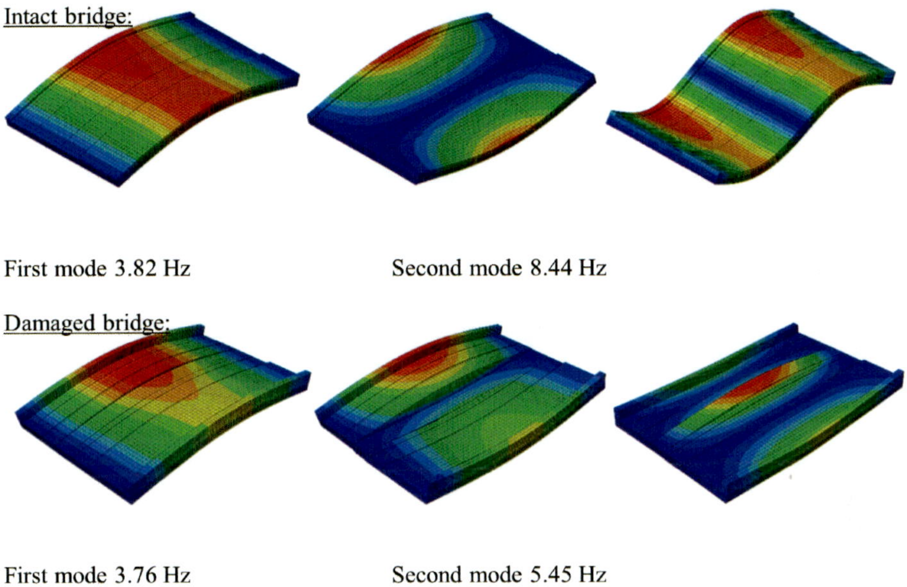


**FIGURE 6.21** Changes in dynamic response (modal shapes) [474]

evaluating and rating the FRP-RSC elements. As a minimum, the detailed description of observations should include specific information such as:

- **Location of observed condition**: Reference the location of the FRP-RSC element from an easily identifiable point when documenting the observed condition. Section 2 of the AASHTO Manual for Bridge Element Inspection [475] gives the inspector a master location matrix of all the elements and identification numbers for quick reference. Specify the location of the deficiency within the element to provide clear and precise references for the location. Consider including a sketch to make it easier to identify the location.
- **Extent of the observed deficiency or condition**: Record the width, length, and depth of the observed condition; include sketches of the condition to supplement the written description.
- **Type of deficiency or observed condition**: This can include observations from several inspection methods. Visual inspection could give an indication of discoloration, cracks, delamination, and wrinkling. In acoustic or tap-test inspections, the types of deficiency could include debonding, delamination, and voids between FRP wraps and underlying concrete.
  - **Severity of observed condition or deficiency**: While the severity level is subjective, it should be determined using a consistent scale. The scale used for rating the severity of a condition should be uniform across all bridge components.

- **Potential reasons for the identified condition or deficiency**: If feasible, make a note of the presence or absence of contributing factors. In case of cracks or other deficiencies, observe if these cracks or defects are active, meaning they are affected by loading or environmental conditions.
- **Photographs, audio recordings, videos, or other documentary evidence of the observed condition at the time of inspection**: The documentary evidence should be indexed with an appropriate description and should be cross-referenced with the field notes.

### 6.7.2 Evaluation of Historic Data

Maintaining a record of historical data is critical to accurately identify trends in the condition of a bridge over time. This data can provide valuable information on bridge maintenance needs, help predict future problems, and guide decision-making on repair and component replacement.

Inspectors should thoroughly examine previous inspection findings related to the following:

- Inspectors review all available quality assurance and quality control data for the materials used for the construction of the reinforced concrete element. If the element has undergone prior strengthening, inspectors should also examine any available information concerning the FRP system used, if applicable.
- Recorded observations taken during and immediately after construction.
- Information on any incidents or circumstances that may have caused deficiency or impacted the condition of the bridge, including vehicular impact, fire damage, or chemical damage.
- Progression of deficiency or deterioration (if any) or the strengthening/repair/retrofit system used to address the incident/circumstance.
- Record whether the circumstances causing the previous deterioration are still present.

Review these factors in all preceding inspection reports and inspect the same areas identified in previous reports during the most recent inspection to monitor the development of these conditions over time. Analyzing the data from earlier reports against the current observations will enable the inspector to assess the gravity of the condition, the likely progression of deficiency, the condition's probable influence on the RC element, and the urgency of remedying any harmful conditions.

### 6.7.3 Standard Checklists

#### 6.7.3.1 Pre-Inspection Checklist

- All necessary personal safety equipment;

- Flashlight;
- Small mallet or hammer;
- Feeler gauges;
- Tap tester;
- Camera;
- Notepad, pencils, etc.;
- Printed forms;
- Detailed drawings of bridge and FRP-RSC member;
- Any available specifications about the materials used for the construction and/or strengthening of the FRP-RSC element;
- Other NDT devices as per the inspection scope;
- Site and operation safety analysis;
- MOT plans and coordination;
- Access and rigging plan and equipment.

#### 6.7.3.2 FRP-RSC Damage Types Checklist

- The checklist of evidence/cause for FRP in internal application (FRP-RC) is as follows:
  i. water seepage/stains
  ii. presence of environmental exposure (water, alkaline, saline exposure),
  iii. cracks/spalling,
  iv. excessive deflections under live load or dead load alone,
  v. other signs of distress
  vi. presence of factors such as elevated temperature, freeze/thaw exposure, etc.
- The checklist of deficiencies for FRP in external application is as follows:
  vii. voids: fiber–matrix interface (intrinsic to composites),
  viii. wrinkling,
  ix. blistering,
  x. fiber–matrix debonding and delamination between composite layers,
  xi. fiber exposure,
  xii. scratches,
  xiii. cracks,
  xiv. discoloration,
  xv. impact damage,
  xvi. voids/debonding between FRP and concrete,
  xvii. delamination/cover separation,
  xviii. cracks in concrete,
  xix. voids in concrete,
  xx. corrosion in steel reinforcement of concrete substrate

#### 6.7.3.3 Inspection and Test Methods Checklist

The inspection and test methods are as follows:

Methods that normally performed within routine inspection:

- Visual Inspection (VT)
- Tap Testing (TT)

Methods that normally performed as special inspections warranted by routine inspection:

- Infrared Thermography Testing (IR)
- Ground-Penetrating Radar (GPR)
- Ultrasonic Testing (UT)
- Impact Echo Testing (IE)
- Microwave Testing (MW)
- Acoustic Emission Testing (AE)
- Laser Testing Method (LT)
- Radiographic Testing (RT)

## 6.8 INSPECTION FORMS

Tables 6.4,6.5,6.6, and 6.7 contain inspection summary forms that serve as a useful tool for organizing the data gathered by inspectors during the inspection of FRP-RSC elements. These forms are designed to help the inspector categorize and rate the observed conditions, with the aid of field notes, sketches, photographs, and other relevant documentation. It is important for the inspector to utilize these forms effectively to ensure an accurate and comprehensive assessment of the bridge's condition. It should be noted that these tables are shown as an example and are subject to modification as needed.

**TABLE 6.4**
**Bridge Inspection Program Information**

| | | | |
|---|---|---|---|
| **Bridge ID/Number:** | | **Inspection Date: / /** | |
| **Bridge Name:** | | **Inspection Type:** | |
| **Bridge Location:** | | **Year Built:** | |
| **FRP Application (External or Internal):** | | **Last Inspection: / /** | |
| **Feature Carried:** | | | |
| **Feature Under:** | | | |
| **Inspection Agency:** | | | |
| **Inspection Team (Circle Team Leader):** | | | |
| Bridge Details | | | |
| i. Total Length | | | |
| ii. Overall Str. Width | | | |
| iii. Structure type (S=simple span, C=continuous, etc.) | | | |
| iv. Number of spans and span lengths | | | |
| Element Details | | | |
| i. Length | | | |
| ii. Width | | | |
| iii. Depth | | | |
| iv. Reinforcement detail (if available) | | | |
| v. Element type (beam, column, deck, etc.) | | | |
| vi. Element ID | | | |
| FRP Details | | | |
| i. Dimension (length, width-dia.) | | | |
| ii. Fiber type (G/C/B/A-FRP) | | | |
| iii. FRP Location | | | |
| Manufacturing Process | | | |
| i. Wet layup | | | |
| ii. Precured laminates | | | |
| iii. Pultrusion | | | |
| iv. Others | | | |
| Material Properties | | | |
| i. FRP modulus of elasticity | | | |
| ii. FRP tensile strength | | | |
| iii. Concrete compressive strength | | | |
| Inspection/Damage Detection Methods Used | Yes? | Detailed field note page reference | Instrument(s) used |
| i. Visual Inspection (VT) | | | |
| ii. Tap Testing (TT) | | | |
| iii. Impact Echo Testing (IE) | | | |
| iv. Ground Penetrating Radar (GPR) | | | |
| v. Ultrasonic Testing (UT) | | | |
| vi. Infrared Thermography Testing (IR) | | | |
| vii. Global Structure Response (GSR) | | | |
| viii. Other: | | | |
| Note: Supplementary drawing attached? | | | |

**TABLE 6.5**
**Element Inspection Summary Form**

| | | | |
|---|---|---|---|
| **Bridge ID/Number:** | | **Inspection Date: / /** | |
| **Bridge Name:** | | **Inspection Type:** | |
| | | **Inspection Technique:** | |
| **Bridge Location:** | | **Year Built:** | |
| **FRP Application (External or Internal):** | | **Last Inspection: / /** | |
| **Feature Carried:** | | | |
| **Feature Under:** | | | |
| **Inspection Agency:** | | | |
| **Inspection Team (Circle Team Leader):** | | | |
| **Element inspected:** | | **Element ID:** | |
| Element Detailing | | | |
| i. Width, Depth and Length | | | |
| ii. Reinforcement (if available) | | | |
| iii. Concrete compressive strength (if available) | | | |
| Deficiency Types Observed Based on Location | Yes? | Detailed field note page reference | Is the deficiency active? (causal element exists) |
| i. **Concrete** | | | |
| Delamination/Cover Separation | | | |
| Voids | | | |
| Cracks | | | |
| Others | | | |
| ii. **Internal FRP Application** | | | |
| Evidence pointing to misplacement or potential deficiency (If yes, describe the evidence below) | | | |
| Evidence | | | |
| Others | | | |
| iii. **External FRP Application** | | | |
| Voids: Fiber–Matrix Interface (Intrinsic to Composites) | | | |
| Wrinkling | | | |
| Blistering | | | |

(*Contiued*)

**TABLE 6.5 (CONTINUED)**
**Element Inspection Summary Form**

| | | | |
|---|---|---|---|
| Fiber-Matrix Debonding & Delamination Between Layers | | | |
| Fiber Exposure | | | |
| Scratches | | | |
| Discoloration | | | |
| Impact Damage | | | |
| Voids/Debonding Between FRP and Concrete | | | |
| Corrosion in Steel Reinforcement (if any) | | | |
| Others | | | |
| **Global Response** | | | |
| Excessive Deflection | | | |
| Change in Static/ Dynamic Response | | | |
| Others | | | |
| Supplementary NDT/ other forms | | | |

**TABLE 6.6**

**Example of GPR Data Collection**

| | | | | |
|---|---|---|---|---|
| **Ground Penetrating Radar (GPR) Inspection Form** | | | | |
| Element ID | | | Inspection Date | |
| **GPR Measuring Presets** | | | | |
| Filename | | | Device Name & Central Frequency | |
| **Measuring Mode:** | | | | |
| ◆Line Scan | | | □◆Grid Scan | |
| Line Number | | | Grid Number | |
| Gain | | | Grid Size | |
| Filter | | □◆On □◆Off | Grid Resolution | |
| Concrete Cal | | | Concrete Cal | |
| Length | | | **Location of Grid Origin** | |
| **Location of Line of Inspection** | | | Distance from 1st Reference Edge | |
| Distance from Reference Edge (DE) | | | Distance from 2nd Reference edge | |
| Edge Reference No. | | | Edge Reference No. | 1st Edge: |
| | | | | 2nd Edge: |
| Picture No. | | | Picture No. | |
| **Interpretations** | | | | |
| **Element** | | Backwall detectable? □◆Yes □◆No | Detected depth of backwall: | Remarks: |
| **Internal FRP Application** | Rebar ID | | | |
| | Size (if discernible) | | | |
| | Location/Spacing | | | |
| | Detectability | | | |
| | Discontinuity | | | |
| **Other Remarks** | | | | |

**TABLE 6.7**

**Example of UT Data Collection**

| **Phased Array Ultrasonic (PAU) Inspection Form** | | | |
|---|---|---|---|
| Element ID | | Inspection Date | |
| **PAU Measuring Presets** | | | |
| File name | | Device Name | |
| **Measuring Mode:** | | | |
| □ Line Scan | | □◆Area Scan | |
| X spacing (m) | | Y spacing (m) | |
| Analog Gain (dB) | | Analog Gain (dB) | |
| Time Gain Compensation (dB) | | Time Gain Compensation (dB) | |
| Pulse Delay (ms) | | Pulse Delay (ms) | |
| Maximum Transmission Time | µs | Maximum Transmission Time | µs |
| **Location of Line of Inspection** | | | |
| Distance from Reference Edge (DE) | | Distance from Reference Edge (DE) | |
| Edge Reference No. | | Edge Reference No. | |
| Picture No. | | Picture No. | |
| **Calibration** | | | |
| Raw Data Offset (µs) | | Raw Data Offset (µs) | |
| Global Pulse Velocity | m/s | Global Pulse Velocity | m/s |
| **Image Processing** | | | |
| Digital Gain (dB) | | Digital Gain (dB) | |
| Surface Wave Cancellation | □◆On □◆Off | Surface Wave Cancellation | □◆On □◆Off |
| **Interpretations** | | | |

*(Continued)*

**TABLE 6.7 (CONTINUED)**
**Example of UT Data Collection**

| Element | Backwall detectable? □◆Yes □◆No | Detected depth of backwall: | Remarks: |
|---|---|---|---|
| **Internal Application** | Rebar ID | | |
| | Depth | | |
| | Detectability | | |
| | Discontinuity | | |
| **Other Remarks** | | | |

# 7 Condition Rating

## 7.1 INTRODUCTION

The approach used to assess the condition of FRP-RSC structures in this manual closely aligns with contemporary inspection and rating practices employed for conventional bridge materials. While incorporating some adjustments to accommodate procedural variations that account for the unique FRP properties, the condition-rating framework outlined in this document yields ratings that can seamlessly integrate with both federal and state-specific rating procedures. Accordingly, this section develops a condition rating based on the Specifications for the National Bridge Inventory (SNBI) [476] method for bridge component rating.

## 7.2 SPECIFICATIONS FOR THE NATIONAL BRIDGE INVENTORY (SNBI) CONDITION RATING

According to SNBI [476], condition ratings are used to describe or evaluate an existing bridge or culvert compared with its as-built condition. The SNBI uses a uniform condition rating scale of 0 to 9 to rate the general condition of bridges and culverts, as shown in Table 7.1.

### 7.2.1 SNBI Condition Rating of Bridges with FRP Application

This section discusses the rating protocols for bridges with FRP application based on the SNBI condition-rating scale.

#### 7.2.1.1 Internal Applications: FRP-Reinforced Concrete (FRP-RC)

The condition assessment of bridges reinforced with FRP can be performed by adopting the current SNBI practices used for inspecting and rating bridge components reinforced with conventional steel reinforcements. Hence, the general condition rating table presented in Table 7.1 will also be applicable for the condition rating of FRP-reinforced concrete components. An example of determining the condition rating for the FRP reinforced deck by correlating the field observations with Table 7.1 is given in Table 7.2.

#### 7.2.1.2 External Applications: FRP-Strengthened Concrete (FRP-SC) and Stand-Alone FRP Elements

On the other hand, for the external and stand-alone applications of FRP in bridges, based on the terminologies defined in the SNBI (Table 7.3), the condition-rating

DOI: 10.1201/9781003646969-7

**TABLE 7.1**
**Bridge condition rating SNBI scale [476]**

| Code | Condition | Description |
|---|---|---|
| 9 | Excellent | Isolated inherent defects. |
| 8 | Very good | Some inherent defects. |
| 7 | Good | Some minor defects. |
| 6 | Satisfactory | Widespread minor or isolated moderate defects. |
| 5 | Fair | Some moderate defects; strength and performance of the component are not affected. |
| 4 | Poor | Widespread moderate or isolated major defects; strength and/or performance of the component is affected. |
| 3 | Serious | Major defects; strength and/or performance of the component is seriously affected. Condition typically necessitates more frequent monitoring, load restrictions, and/or corrective actions. |
| 2 | Critical | Major defects; component is severely compromised. Condition typically necessitates frequent monitoring, significant load restrictions, and/or corrective actions in order to keep the bridge open. |
| 1 | Imminent failure | Bridge is closed to traffic due to component condition. Repair or rehabilitation may return the bridge to service. |
| 0 | Failed | Bridge is closed due to component condition and is beyond corrective action. Replacement is required to restore service. |

scale shown in Table 7.4 was developed. The approximate extent of defects mentioned in Table 7.4 was adopted from the NCHRP Report 564 [28].

Examples of determining the condition rating for the FRP strengthened bridge components and stand-alone applications of FRP in bridges by correlating field observations with Table 7.4 are given in Tables 7.5 through 7.7.

## 7.3 AASHTO MANUAL FOR BRIDGE ELEMENT INSPECTION (MBEI)

### 7.3.1 Recommendation of FRP as New Material in MBEI

Over the years, the use of FRP composites has been on the rise due to the advantages provided by this material. FRP composites have less density compared to steel and concrete, about one-fourth the density of steel and about two-thirds that of concrete. Accordingly, their application requires less scaffolding and heavy lifting equipment and hence can be handled by a smaller workforce, and as such provides higher safety. Furthermore, their use as an alternative to steel reinforcement in bridge elements reduces the overall weight of the bridge structure, providing many advantages in their design. This calls for considering FRP as a new material within the MBEI, allowing more accurate inspection and maintenance decision-making that in turn

## TABLE 7.2
## Condition rating of FRP-reinforced deck

Example—FRP-Reinforced Deck Condition Rating

FRP-reinforced concrete (FRP-RC) bridge deck approximately 158′ long × 45.3′ wide with the following noted defects.

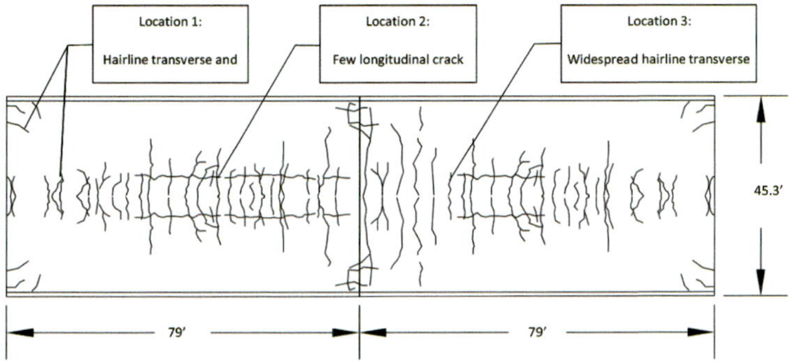


Deck plan view showing defects

Longitudinal cracks on deck surface. Location 2.

Widespread typical transverse hairline cracks with white efflorescence on the soffit of the deck. Location 3.

**Summary of Findings:**

| Location | Defect(s) | Severity | Extent |
|---|---|---|---|
| 1 | Transverse and diagonal cracks | Inherent | Throughout (widespread) |
| 2 | Longitudinal cracks | Minor | A few locations (isolated) |
| 3 | Transverse cracks with efflorescence on the bottom surface of the deck | Minor | Throughout (widespread) |

**Results:** There are areas of isolated/widespread inherent/minor defects that do not affect the strength or performance. Therefore, the superstructure is best characterized as "widespread minor defects". Report 6.

**TABLE 7.3**
**Terminologies adapted from SNBI for developing the condition-rating scale for the FRP application in bridges [476]**

| Terminology | Definition | Remarks |
|---|---|---|
| Defect | A problem with the bridge component that may be caused by deterioration, damage, or an inherent defect. | |
| Inherent defect | Does not necessarily signify damage or deterioration; rather, it is a trait typical of the material or results from normal construction practices. | |
| Minor defect | Damage or deterioration has initiated but is not yet considered significant. | Represents defect severity |
| Moderate defect | Damage or deterioration are significant, but the strength and performance of the component are not affected. | |
| Major defect | Affects the strength and/or performance of the component, as determined by a structural and/or hydraulic review. | |
| Widespread defect | Present in many separate areas of the component. | Represents defect extent |
| Isolated defect | Occurs in one or a few concentrated locations. | |
| Some defect | The defect prevalence is more than isolated and less than widespread. | |

would facilitate the wider use of the material in bridges and buildings. The following discussion is to support this inclusion.

The properties of FRP can be customized as desired by changing the direction and quantity of fibers. Fabricators and engineers have flexibility in the choice of materials. Various types of fiber (glass, carbon, basalt, etc.) and several arrangements (individual fibers, chopped strand mat, woven or stitched fabrics of various geometries) can be used for FRP composites. This allows fabricators to use the FRP materials efficiently and offers composites tailored to the customers' needs [38–41].

The research shows that FRP materials have low degradation in the long term [42] and are good at fatigue resistance. Also, FRP composites have better resistance to salt, water, and other chemicals compared to conventional materials. Oil as well as other heavy hydrocarbons cannot affect FRP composites as much as they can affect conventional materials [43–46].

Although FRP has provided many advantages, there are some downsides to its use. Firstly, the cost of FRP is higher than that of conventional materials, affecting the overall cost of a project. However, this can be seen as an incentive to employ it in an efficient, innovative, and cost-effective way. Also, the initial cost of GFRP rebars has benefited in recent years due to price fluctuation in the metal market worldwide

**TABLE 7.4**

**Condition rating of the external and stand-alone applications of FRP in bridges**

| Code | Condition | Description |
|---|---|---|
| 9 | Excellent | Excellent condition with isolated inherent defects, typically new construction. |
| 8 | Very good | Some inherent defects, with no significant problems noted. |
| 7 | Good | Some minor defects in the form of hairline cracks on the surface with no signs of delamination, debonding, voids, wrinkling, discoloration, blisters, and fiber exposure. |
| 6 | Satisfactory | Isolated moderate defects in the form of shallow cracks, scratches, abrasion, blistering, delamination, and debonding over less than 2% of surface area total. Fibers are not exposed, ruptured, or buckled at the surface damage locations. Delamination smaller in every dimension than 4 in. and away from structural details or located such that structural function will not be impaired. |
| 5 | Fair | Some moderate defects in the form of shallow cracks, scratches, abrasion, blistering, delamination, and debonding extends over 2% to 10% of surface area total. Fibers exposed but not ruptured, buckled, or debonded at the surface damage locations. Delamination smaller in every dimension than 6 in. and away from structural details or located such that structural function will not be impaired. |
| 4 | Poor | Widespread moderate defects such as surface damage in the form of cracks, scratches, abrasion, blistering, delamination, and debonding extends over 10% to 25% of the total area. Fibers in the cracks exposed but not debonded, buckled, or ruptured at the surface damage locations. Delamination smaller in every dimension than 6 in. but near structural details or located such that to have structural effects. |
| 3 | Serious | Some major defects such as surface damage in the form of deep cracks, scratches, abrasion, blistering, delamination, and debonding extends over more than 25% of area total. Fibers are visibly exposed and debonded, but not ruptured or buckled at the surface damage locations. Delamination smaller in every dimension than 14 in. Structural analysis may be required to determine the functionality of the strengthened element. |
| 2 | Critical | Widespread major defects such as fibers are exposed, debonded, and ruptured, or buckled at the surface damage locations, and deep cracks, especially if perpendicular to major stress directions but not through rupture or puncture. Delamination larger in any dimension than 24 in. Condition typically necessitates frequent monitoring, significant load restrictions, and/or corrective actions. |
| 1 | Imminent failure | The condition warrants a structural review to determine the effect on strength or serviceability of the element or bridge; OR a structural review has been completed and the defects impact strength or serviceability of the element. Repair or rehabilitation may return the bridge to service. |
| 0 | Failed | Out of service—beyond corrective action. |

## TABLE 7.5
## Condition rating of FRP-strengthened superstructure

Example—FRP-Strengthened Superstructure Condition Rating

Single span adjacent box girder bridge. Each box girder is 3 ft. wide, 27 in. deep, and 54 ft. long. The following defects are noted.

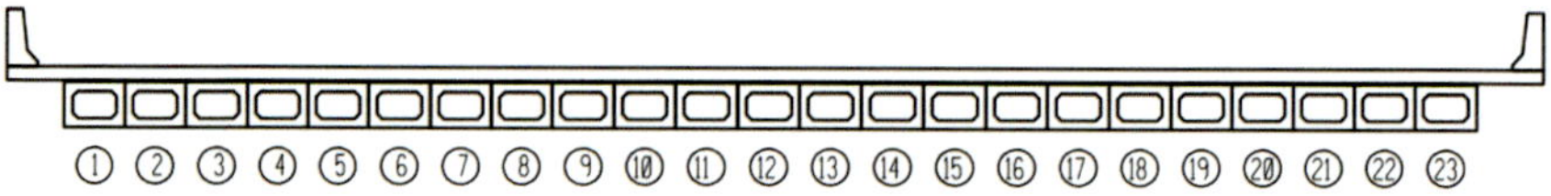


Cross-section view, looking south, indicating box girder numbering

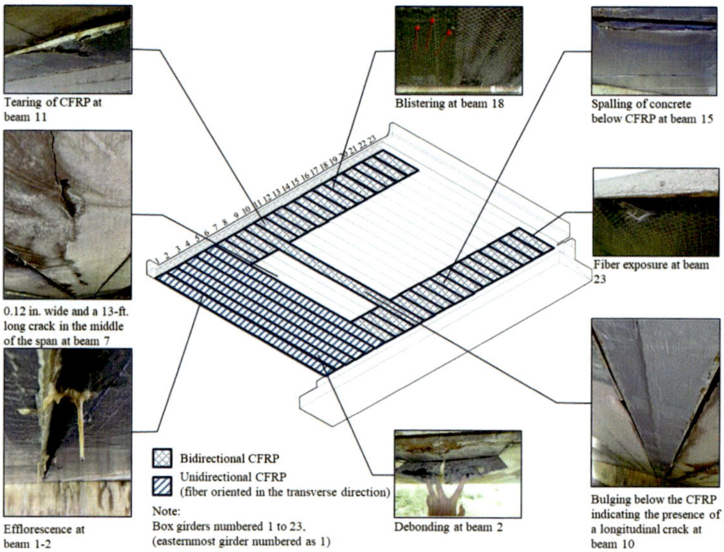


3D view of the box beams showing superstructure defect locations. Source: [477]

**Summary of Findings:**

| Location | Defect(s) | Severity | Extent |
|---|---|---|---|
| Beam 18 | Blistering | Minor | One location (isolated) |
| Beam 23 | Fiber exposure | Minor | One location (isolated) |
| Beam 2 | Debonding | Major | More than a few locations (some) |
| Beam 11 | Crack or tearing of CFRP | Major | One location (isolated) |
| Beam 1-2 | Discoloration or stains | Minor | A few locations (isolated) |
| Beam 15 | Substrate spalling | Major | One location (isolated) |
| Beam 10 | Substrate crack | Major | One location (isolated) |
| Beam 7 | Crack in unstrengthened concrete | Major | A few locations (isolated) |

**Results:** There are several instances of isolated/some major defects that can be characterized together as "some major defects" and isolated minor defects that do not affect the strength or performance. Therefore, the superstructure is best characterized as "some major defects". Report 3.

## TABLE 7.6
## Condition rating of FRP-strengthened substructure

Example—FRP-Strengthened Substructure Condition Rating

Continuous steel girder bridge. The bents were reinforced concrete with a crashwall and rectangular (30″ × 36″) columns. The crashwall was wrapped with GFRP laminates, with fibers running parallel to the longitudinal axis of the column. Columns were completely wrapped on all four sides with fibers oriented perpendicular to the longitudinal axis of the column. The columns were wrapped approximately 10 ft up measured from the top of the crashwall. The following defects are noted in Bent 5.

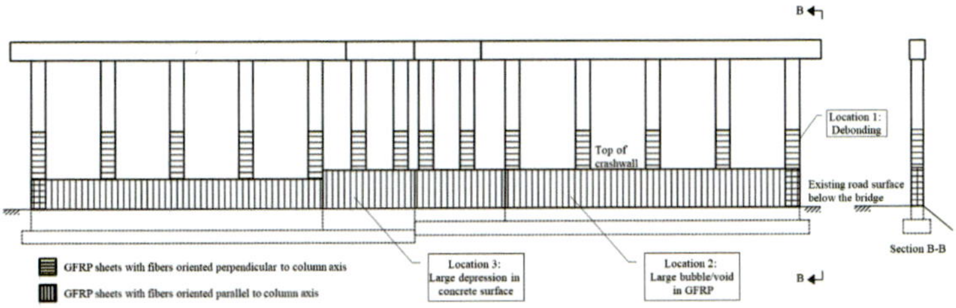


Elevation view of two concrete column piers showing substructure defect locations

Debonding at the easternmost column of Bent 5. Location 1.

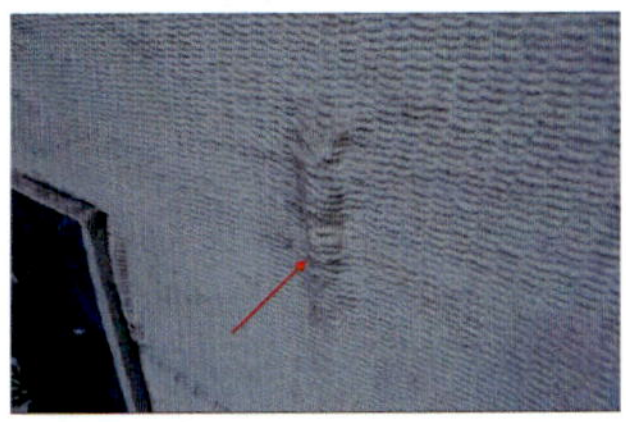

Large bubble/void in GFRP laminate on crashwall of Bent 5. Location 2.

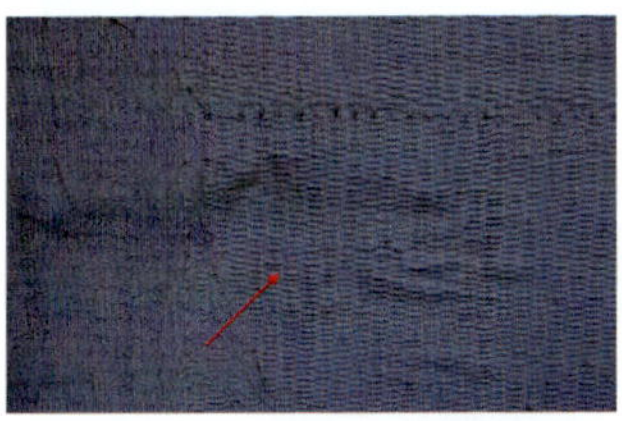

Large depression in concrete surface at crashwall of Bent 5. Location 3.

Source: [470]

**Summary of Findings:**

| Location | Defect(s) | Severity | Extent |
|---|---|---|---|
| 1 | Debonding | Moderate | More than a few locations (some) |
| 2 | Blister/Bubble | Minor | One location (isolated) |
| 3 | Void | Minor | More than a few locations (some) |

**Results:** There are instances of some minor/moderate defects that can be characterized together as "some moderate defects". Therefore, the substructure is best characterized as "some moderate defects". Report 5.

**TABLE 7.7**

**Condition rating of FRP deck (stand-alone FRP application)**

Example—FRP Deck Condition Rating

Five-span bridge with FRP deck supported over six built-up steel stringers. The following defects are noted.

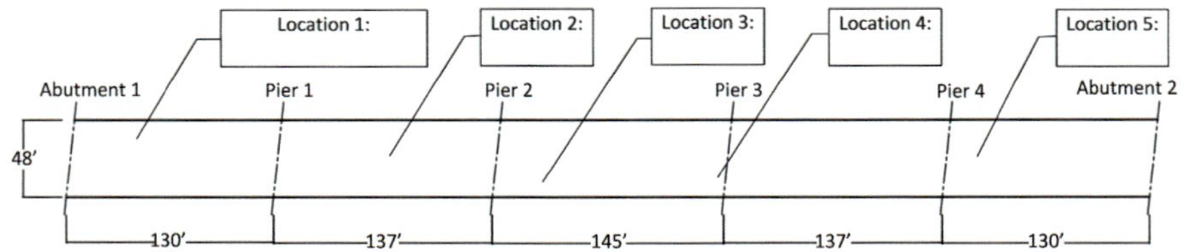


Deck plan view showing defects

Wearing surface delamination. Location 1.

Blisters. Location 2.

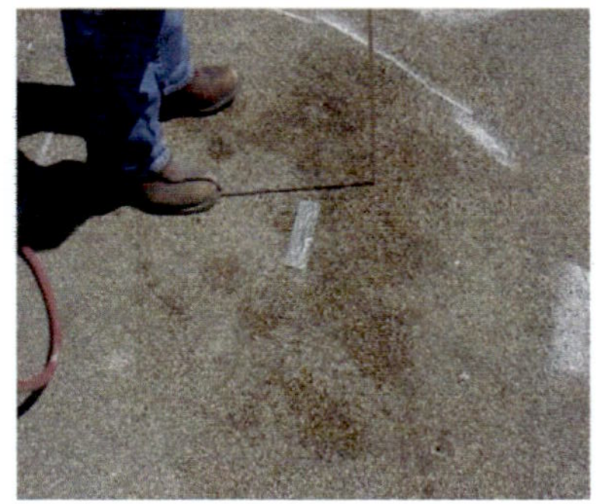

Water accumulation. Location 3.

(*Contiued*)

**TABLE 7.7 (CONTINUED)**

**Condition rating of FRP deck (stand-alone FRP application)**

Debonding (detected using tap testing). Location 4.

Cracks and spalling in wearing surface. Location 5.

Source: [28]

**Summary of Findings:**

| Location | Defect(s) | Severity | Extent |
|---|---|---|---|
| 5 | Cracks and spalling in wearing surface | Moderate | More than a few locations (some) |
| 1 | Wearing surface defect | Minor | One location (isolated) |
| 2 | Blisters | Minor | One location (isolated) |
| 3 | Discoloration/water accumulation | Minor | One location (isolated) |
| 4 | Debonding | Minor | One location (isolated) |

**Results:** There are several instances of isolated minor/moderate defects that can be characterized together as "some minor defects". Therefore, the substructure is best characterized as "some minor defects". Report 7.

since the mid-2020s and has decreased due to the growth of the GFRP rebar market [478]. And when the long-term costs are included for bridge maintenance purposes, the cost difference will narrow or turn the other way.

Secondly, FRP has a low coefficient of thermal expansion; some CFRP composites have a negative coefficient. This can be a problem when the FRP is used in conjunction with conventional materials such as concrete. Also, the resin properties are temperature-dependent. Lower temperatures lead to an increase in strength and make the resin more brittle. Furthermore, with the increase in temperature, a critical point, which is the glass transition temperature of a resin, may be reached and cause a rapid decrease in strength. This issue can be prevented by choosing a resin type whose glass transition is higher than the anticipated maximum operating temperature [38]. The proprietary aspect of FRP material may also be viewed as a limitation. The bendability, or lack of it, has also been considered a major issue; however, this problem is being addressed by using thermoplastic resins. As a result, FRP bars can be bent at the reinforcement fabricator instead of at the pultruder, following the same model as steel rebars [132].

Considering all the above-mentioned advantages and the results of the literature review reported earlier that point to extensive and increasing use of FRP in bridge structures in different forms, it is recommended that FRP be included as a new material in the Manual for Bridge Element Inspection (MBEI) [475]. To be able to provide element condition severity, one should first recognize the bridge elements that use FRP; this is accomplished in the following section.

### 7.3.2 Identifying Bridge Elements Using FRP Externally

Based on the literature review, Figure 7.1 shows the summary of the applications of FRP material in bridge construction. These are categorized as internal application in reinforced concrete, external applications for concrete elements, and stand-alone applications where the FRP composites act as structural or non-structural elements. However, it should be noted that the bridge elements with internal application of FRP are mostly included under sections "3.1.1—Reinforced Concrete" and "3.1.2—Prestressed Concrete" in the Manual for Bridge Element Inspection (MBEI) [475]. Therefore, the bridge elements that can be constructed using FRP as the primary material, as well as those bridge elements that use FRP externally for strengthening, repair, and rehabilitation, were identified for addition to the relevant sections of the Manual for Bridge Element Inspection (MBEI) [475]. These bridge elements fall under the category of external application and stand-alone application of FRP. Figure 7.1 illustrates those elements with exposed FRP, either constructed mainly using FRP composite or with externally applied FRP under items (h) through (t).

#### 7.3.2.1 Identifying Damage and Defects

Detailed description of damage and defects in FRP can be found in previous chapters.

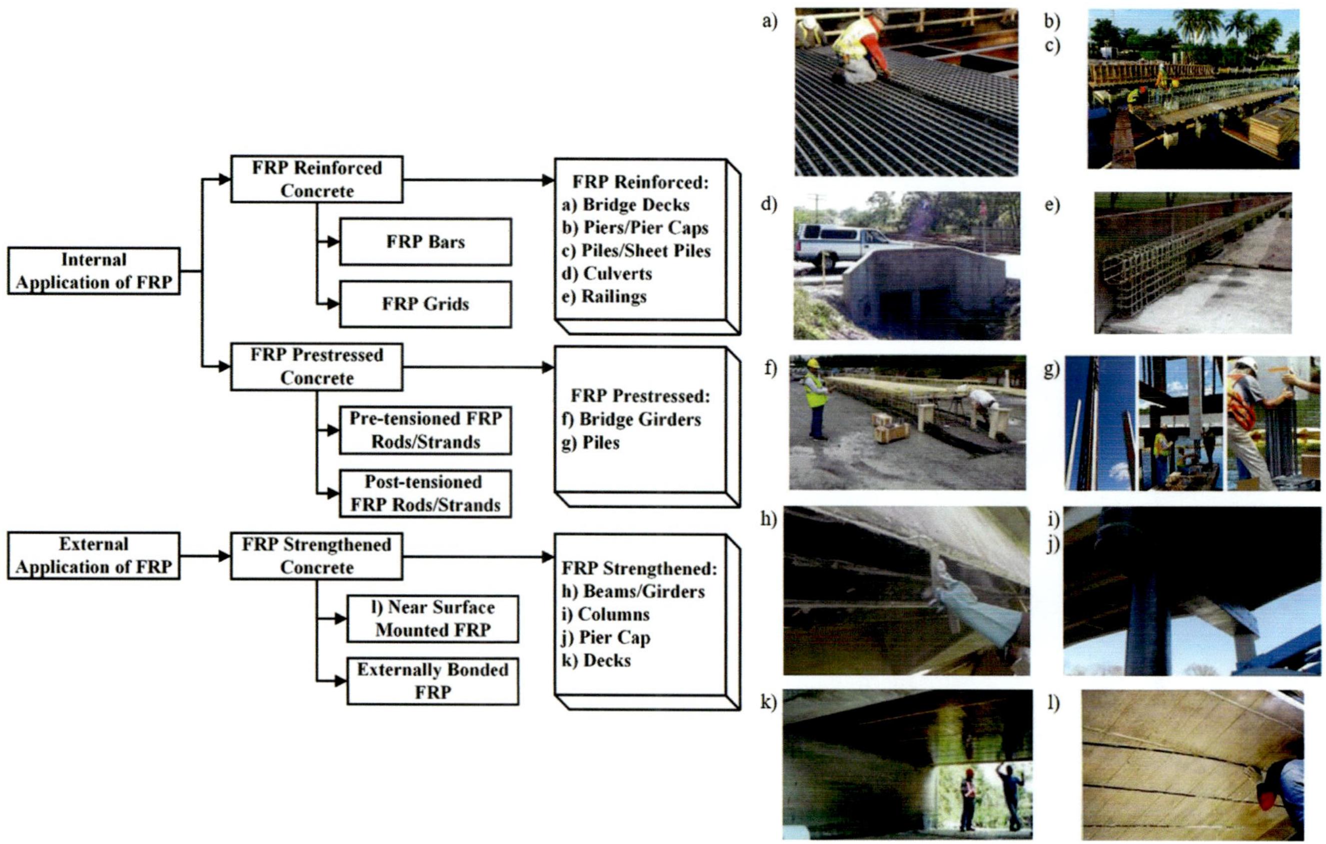


**FIGURE 7.1** Applications of FRP material in bridge construction

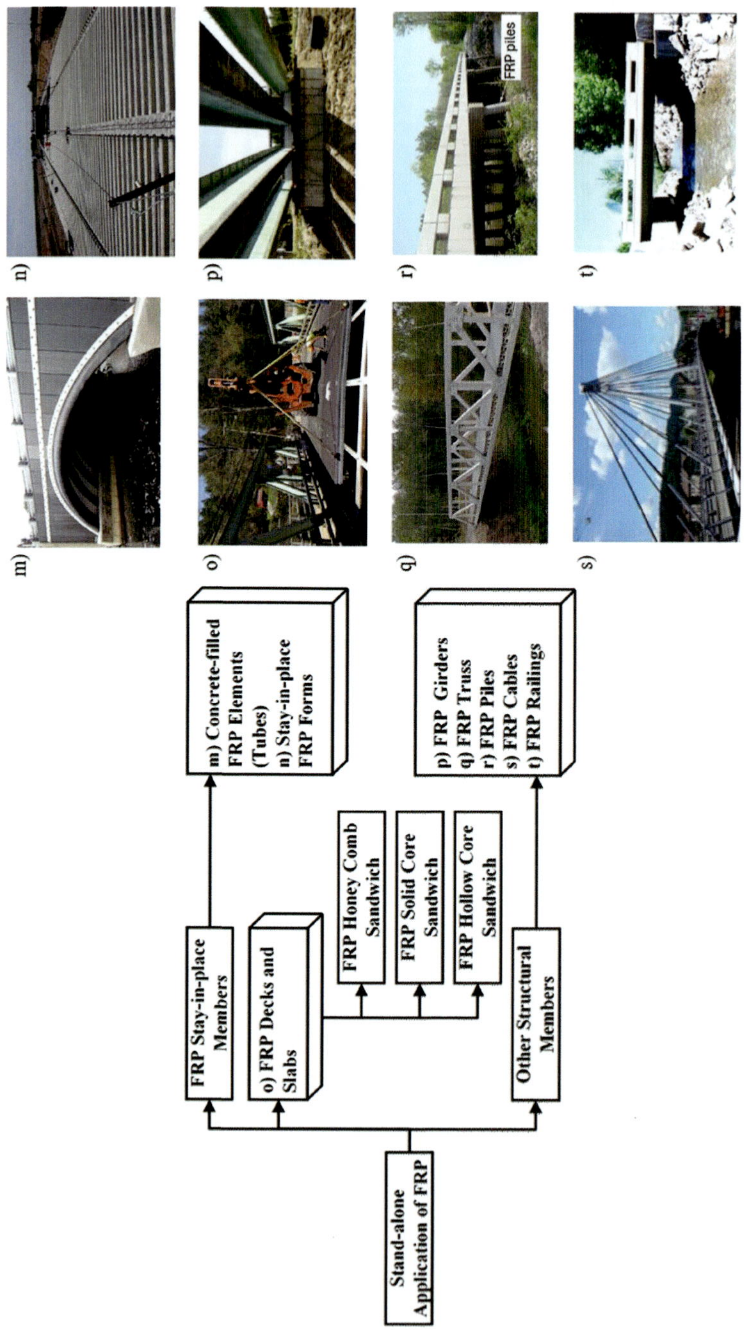


**FIGURE 7.1** Continued.

### 7.3.2.2 Guide for Condition State Determination

Based on the type of damage/defects in FRP as described earlier, their severity and consequence for structural performance, the following table was developed to identify the condition state as per MBEI.

| Defects | Condition States | | | |
|---|---|---|---|---|
| | CS 1 | CS 2 | CS 3 | CS 4 |
| | GOOD | FAIR | POOR | SEVERE |
| Defects for FRP-Strengthened Elements and FRP deck and slabs | | | | |
| Cracking (1230)<br>Scratches (1240)<br>Abrasion (1290) | None | Shallow cracks, scratches, or abrasion, limited to the polymer, with no fibers visible and parallel to the major stress direction. | Wide or deep cracks or scratches or abrasion, especially if perpendicular to major stress or fiber directions but no through rupture or puncture. | The condition warrants a structural review to determine the effect on strength or serviceability of the element or bridge; OR a structural review has been completed and the defects impact strength or serviceability of the element. |
| Wrinkling (1210)<br>Voids (1270)<br>Blister (1260) | None | Less than 1 inch raised or less than 6 inches in diameter or length. | More than 1 inch raised or more than 6 inches in diameter or length. | |
| Discoloration (1250) and Fire Damage (1215) | None, or cosmetic stain or fire mark | Permanent shallow discoloration or UV exposure and protective coating degradation 6 inches in diameter or less. | Permanent discoloration or UV exposure with cracking, brittleness, or distress, or more than 6 inches in diameter. | |
| Fiber Exposure (1280) | None | Exposed but not ruptured, buckled, or debonded at the surface damage location. | Fibers exposed and debonded (especially in cracks perpendicular to major stress or fiber directions) but not buckled or ruptured at the surface damage location. | |

*(Contiued)*

| Defects | Condition States | | | |
|---|---|---|---|---|
| | CS 1 | CS 2 | CS 3 | CS 4 |
| | GOOD | FAIR | POOR | SEVERE |
| Delamination/ debonding (interlaminar/ from substrate) (1205) | None | Delamination smaller in every dimension than 6 inches but away from sensitive details (e.g., connections) to have structural effects. | Delamination smaller in every dimension than 6 inches but near sensitive details (e.g., connections) to have structural effects. | |
| Defects applicable only to FRP deck and slab (adapted from NCHRP report 564) | | | | |
| Panel-to-Panel Joint (1225)<br>Panel-to-Girder Joint (1235)<br>Approach Joint (1245) | None | Evidence of joint degradation visible due to cracking in overlay or topping above panel joint locations. Gaps and cracks of up to 1/16 inch; no loss of bolts, clips, or other devices. Leakage may be minimal. | Gaps, misalignment, and cracks of up to ¼ inch in overlay. Few clips or bolts loose or lost, minor elevation changes for adjacent panels evident. Leakage may be moderate, more than a drip and less than a free flow of water. | |
| Facesheet Debonding (1255) | None | Visible minor bulge or peeling on the top surface indicating initiation of debonding. | Noncritical damage to surrounding structural elements, such as reflective cracking at wearing surface or local damage to joints resulting in detection of debonding. | |

*(Contiued)*

| Defects | Condition States | | | |
|---|---|---|---|---|
| | CS 1 | CS 2 | CS 3 | CS 4 |
| | GOOD | FAIR | POOR | SEVERE |
| General | | | | |
| Settlement (4000) | None | Exists within tolerable limits or arrested with no observed structural distress. | Exceeds tolerable limits but does not warrant structural review. | |
| Scour (6000) | None | Exists within tolerable limits or has been arrested with effective countermeasures. | Exceeds tolerable limits but is less than the critical limits determined by scour evaluation and does not warrant structural review. | |
| Damage (7000) | Not applicable | The element has impact damage. The specific damage caused by the impact has been captured in CS 2 under the appropriate material defect entry. | The element has impact damage. The specific damage caused by the impact has been captured in CS 3 under the appropriate material defect entry. | The element has impact damage. The specific damage caused by the impact has been captured in Condition State 4 under the appropriate material defect entry. |

*Note:* This table is provided as a general suggestion for recognizing the condition state. The inspector may use engineering judgment to recognize exceptions. If any damage, regardless of its extent, is perceived to affect the structural integrity of the element, it should be noted as State 4 or severe.

### 7.3.2.3 Visual Aids for Condition State Determination

Following tables are provided to help determine the condition state for each type of damage/defect, aided by a photograph of actual damage.

**Defect 1230 Cracking (FRP)**

| Condition State 1 | Condition State 2 | Condition State 3 |
|---|---|---|
| Hairline surface cracks | Shallow cracks limited to the polymer, with no fibers visible and parallel to the major stress direction. | Wide or deep cracks, especially if perpendicular to major stress or fiber directions but no through rupture or puncture. |
| 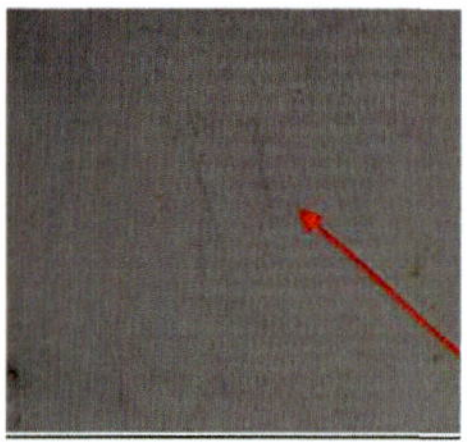 Source: Christopher Williams |  |  |

**Defect 1240 Scratches (FRP)**

| Condition State 1 | Condition State 2 | Condition State 3 |
|---|---|---|
| None | Shallow scratches limited to the polymer, with no fibers visible and parallel to the major stress direction. | Wide or deep scratches, especially if perpendicular to major stress or fiber directions but no through rupture or puncture. |
| | 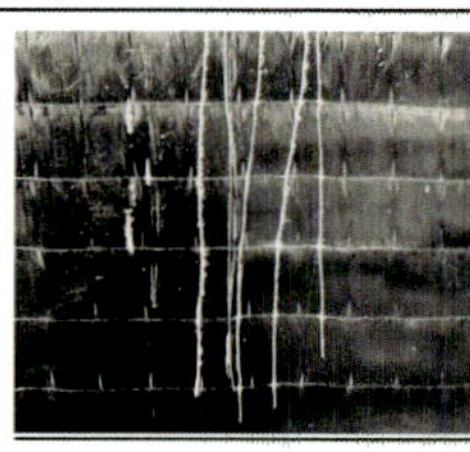 | 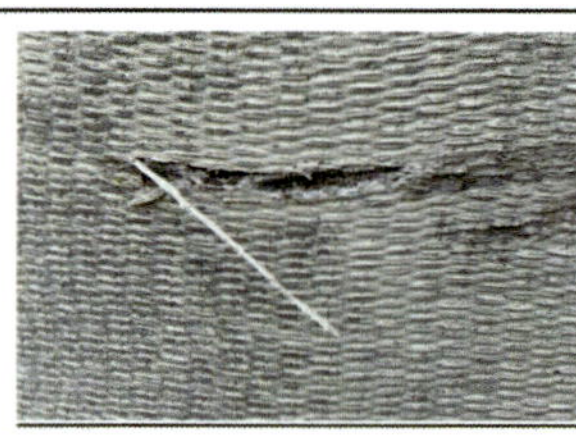 |

## Defect 1290 Abrasion (FRP)

| Condition State 1 | Condition State 2 | Condition State 3 |
|---|---|---|
| None | Shallow abrasion, limited to the polymer, with no fibers visible and parallel to the major stress direction. | Wide or deep abrasion, especially if perpendicular to major stress or fiber directions but no through rupture or puncture. |
| | 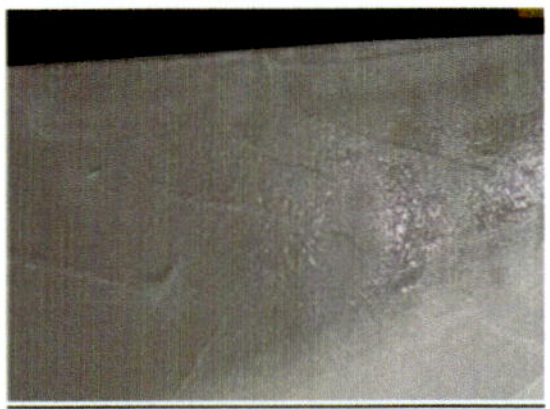 |  |

## Defect 1210 Wrinkling (FRP)

| Condition State 1 | Condition State 2 | Condition State 3 |
|---|---|---|
| None | Less than 1 in. raised or less than 6 inches diameter or length. | More than 1 in. raised or more than 6 inches in diameter or length. |
| |  |  |

### Defect 1250 Discoloration (FRP)

| Condition State 1 | Condition State 2 | Condition State 3 |
|---|---|---|
| None or none after cleaning superficial stain | Permanent shallow discoloration or UV exposure and protective coating degradation of 6 inches in diameter or less. | Permanent discoloration or UV exposure with cracking, brittleness, or distress, or more than 6 inches in diameter. |
| |  |  Source: Christopher Williams |

### Defect 1260 Blister (FRP)

| Condition State 1 | Condition State 2 | Condition State 3 |
|---|---|---|
| None | Less than 1 inch raised or less than 6 inches in diameter or length. | More than 1 inch raised or more than 6 inches in diameter or length. |
| |  |  |

## Defect 1270 Voids (FRP)

| Condition State 1 | Condition State 2 | Condition State 3 |
|---|---|---|
| None | Less than 1 inch raised or less than 6 inches in diameter or length. | More than 1 inch raised or more than 6 inches in diameter or length. |
| | 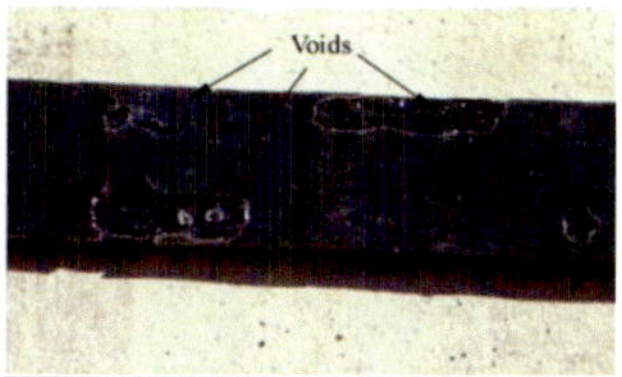 <br>Source: Vistasp Karbhari |  |

## Defect 1280 Fiber Exposure (FRP)

| Condition State 1 | Condition State 2 | Condition State 3 |
|---|---|---|
| None | Exposed but not ruptured, buckled, or debonded at the surface damage location. | Fibers exposed and debonded (especially in cracks perpendicular to major stress or fiber directions), but not buckled or ruptured at the surface damage location. |
| |  |  |

## Defect 1205 Delamination/Debonding (FRP)

| Condition State 1 | Condition State 2 | Condition State 3 |
|---|---|---|
| None | Delamination smaller in every dimension than 6 inches but away from sensitive details (e.g., connections) to have structural effects. | Delamination smaller in every dimension than 6 inches but near sensitive details (e.g., connections) to have structural effects. |
| | 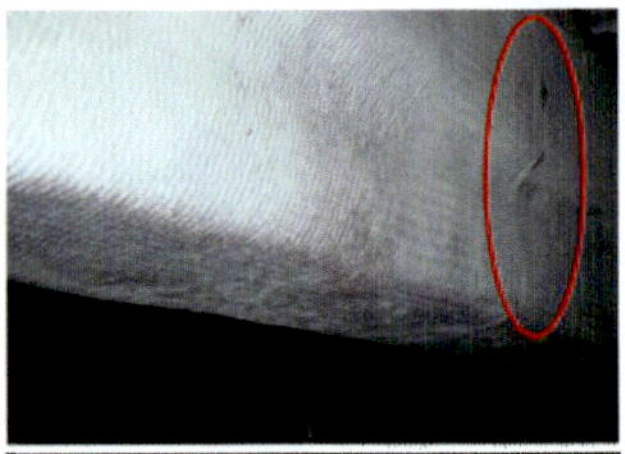 | 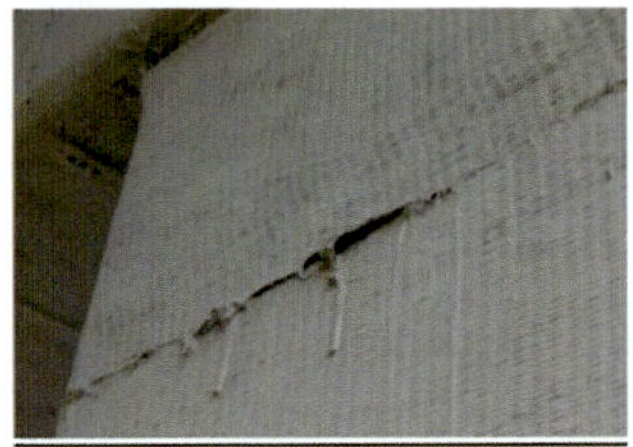 Source: Christopher Williams |

## Defect 1215 Fire Damage (FRP)

| Condition State 1 | Condition State 2 | Condition State 3 |
|---|---|---|
| None or none after cleaning superficial stain or fire mark | Permanent shallow discoloration 6 inches in diameter or less. | Permanent discoloration and with cracking, brittleness, or distress or more than 6 inches in diameter. |
| |  |  |

## Defect 1225, 1235, 1245 Panel Joints (FRP)

| Condition State 1 | Condition State 2 | Condition State 3 |
|---|---|---|
| Minor deterioration with hairline cracks less than 1/32 in. No noticeable water leakage observed from underside. | Evidence of joint degradation visible due to cracking in overlay or topping above panel joint locations. Gaps and cracks of up to 1/16 inch; no loss of bolts, clips, or other devices. Signs of water leakage maybe minimal. | Gaps, misalignment, and cracks of up to ¼ inch in overlay. Few clips or bolts loose or lost, minor elevation changes for adjacent panels evident. Leakage may be moderate, more than a drip and less than a free flow of water. |
|  |  | 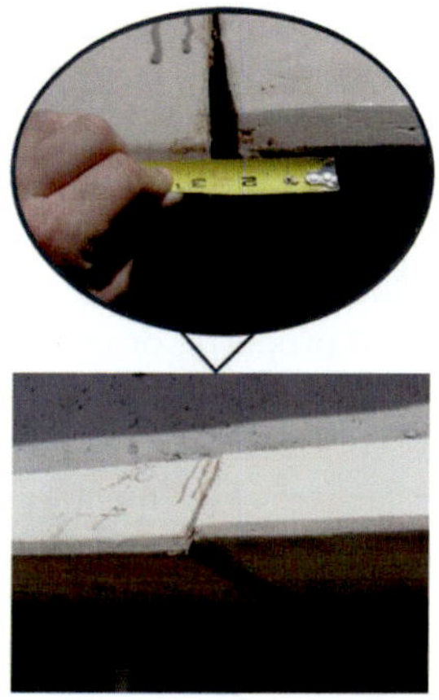 |

## Defect 1255 Facesheet Debonding (FRP)

| Condition State 1 | Condition State 2 | Condition State 3 |
|---|---|---|
| None | Visible minor bulge or peeling on the top surface indicating initiation of debonding. | Noncritical damage to surrounding structural elements, such as reflective cracking at wearing surface or local damage to joints resulting in detection of debonding. |
| |  |  |

## 7.4 CONDITION RATING AS INPUT FOR LOAD RATING

Bridge load rating is defined as the process of determining the safe load-carrying capacity of a bridge, considering its current structural, material, loading, and traffic conditions. The rating process takes into account the original design specifications of the bridge, any deterioration or damage the bridge has sustained over time, and any repairs that have been made. The purpose of a bridge load rating is to ensure the safety and reliability of the bridge by determining if it can safely carry its intended loads, to identify bridges that may need repair, reinforcement, or load restrictions, and to help manage and prioritize bridge maintenance, repair, and replacement programs.

As mentioned earlier, the existing structural conditions, material properties, loading, and traffic conditions influence the load rating of a bridge. Among these factors, the incorporation of FRP material condition in the load rating process would be an additional factor to be considered for FRP-reinforced/strengthened concrete elements. In a survey conducted by Phares et al. [479], it was found that the State DOTs considered the deterioration or damage at the individual bridge element level or at the global level or both by reducing the condition factor, cross-sectional property, material property, or combination of these factors. For instance, the section loss in a steel beam bridge can be accounted for by reducing the thickness of the section. In RC bridges, the loss of capacity due to spalling or delamination along the bottom steel reinforcement can be incorporated through a reduction in the area of the reinforcement, and the deterioration of concrete in the compression zone can be considered through the reduction of compressive strength of concrete. And, in PC beam bridges, the effect of prestress loss and exposed strands can be modeled by excluding those strands and the adjacent strands [479]. However, there is a lack of standardized procedures for accommodating deterioration in the load rating process, specifically in the case of FRP applications in bridges. In most part, it remains the responsibility of the structural engineer in charge of load rating to make inferences and analyze the effect of the conditions reported by the inspection of the structure. There is therefore a demand for further research into the effect of deterioration and damage on load rating that is outside the scope of this study. However, in the context of providing guidance for structural engineers on load rating of such structures, Section 7.4.1 focuses on the approach of inferring and interpreting inspection results as input for load rating of bridges with FRP/RSC elements. Additionally, Section 7.4.2 gives a brief introduction to the AASHTO Manual for Bridge Evaluation (MBE) [480], which is incorporated by reference in the National Bridge Inspection Standards, 23 CFR 650.313(k)(1), [481] to rate each bridge as to its safe load capacity. Finally, Section 7.4.3 introduces several research works conducted in the past which describe the relationship between the deterioration or damages in FRP materials and their performance. Further, it reviews some considerations for long-term effects of damages and remaining life predictions.

### 7.4.1 Inferring and Interpreting Inspection Results as Input for Load Rating of Bridges with FRP/RSC Elements

The uncertainties associated with the nominal resistance of bridge members increase significantly as the member begins to deteriorate from its pristine condition. In addition, once the deterioration has started, the rate of deterioration may increase rapidly, which in turn will further increase uncertainties and resistance variabilities. Uncertainties in determining the true effect of deterioration can be reduced through improved inspection and field measurements [475].

Inferring and interpreting inspection results is crucial for effectively using inspection data to ensure the safety and serviceability of inspected components, and consequently to establish a load rating for bridges. In the case of FRP-RSC, once the damage or defects are identified and categorized within the condition state and rating by the bridge inspectors, they will have to be evaluated and assessed for consideration in the load rating of the bridge by the bridge engineer in charge of load rating based on the input provided by the inspection. The research team believes that for applications such as those containing stand-alone FRP elements and FRP-RC elements, the engineer in charge of load rating can perform the assessment using guidelines currently available for conventional bridge components combined with provision for updating FRP strength as per the inspection results. For applications that include FRP-SC elements, where the element evaluation depends not only on the FRP strengthening but also on the steel-reinforced concrete (RC) substrate and elements, the engineer can analyze the combined system with the use of provisions for updating strengths for the RC and FRP. In either case, the strength of FRP should be estimated based on the input from the inspection results.

Quantifying the somewhat subjective inspection data in a form of strength reduction for FRP/RSC elements is not an easy task and requires sound engineering judgment and analytical skills from the bridge engineer in charge of load rating. The aerospace and defense industries have developed methods to quantify some of these results. These methods involve assessing the severity and risk of failure associated with damage, based on its type and extent, using a standardized scale in military and aerospace settings. For instance, a chart derived from such research [482], shown in Figure 7.2, facilitates the correlation between the damage magnitude (such as hole diameter or delamination area) and the reduced load-bearing capacity of the components. This enables a structured approach to assessing damage in terms of its impact on component strength. While the figure is specifically tailored for aircraft structures, it demonstrates the existence of a feasible and systematic method for analyzing FRP field inspection data. However, it is important to recognize that these aids, found in military handbooks and aerospace references, are the culmination of over 50 years of research, testing, and development. Consequently, it is unlikely that similar aids could be created for the bridge engineering sector within the short timeframe and limited scope of this project. Moreover, aircraft structures undergo inspections at a significantly higher frequency. Nonetheless, it is conceivable to adopt this concept from the defense sector to set a long-term objective of developing analogous aids for the bridge industry.

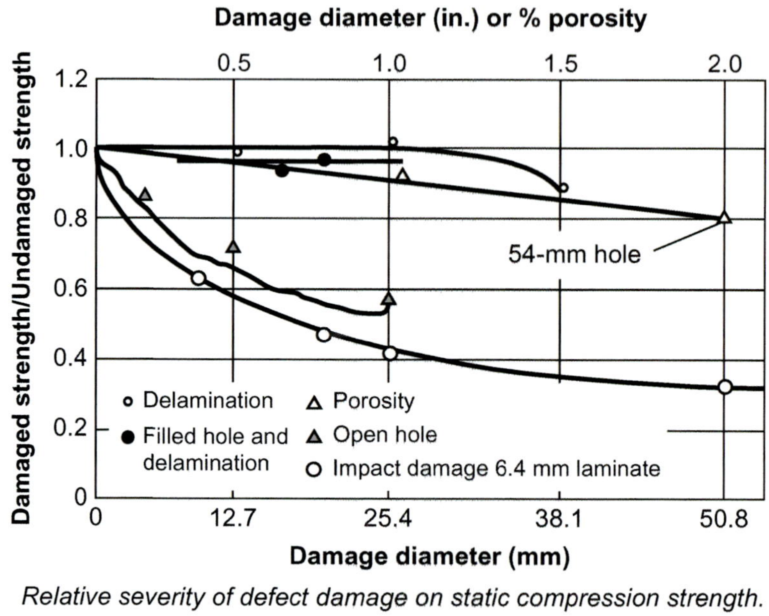


**FIGURE 7.2** Damage versus strength reduction [482]

The research team has developed a framework for inspection that would assist an inspector in evaluating and assessing the condition of the FRP/RSC elements and components. The correlation between the condition and strength could be developed on a case-by-case basis by the engineer in charge of load rating or could be performed based on a future set of provisions (e.g., charts and tables) using some of the test data from aerospace applications or future test results of bridge elements. For example, for a specific type of FRP strengthening, the extent of damage quantified by the delaminated area of FRP could be related to a specific percent reduction in the strength of the SC element. The scale can be set such that at one end, for no delamination, there would be no reduction in the extra load-carrying capacity provided by the FRP presence, while a delamination diameter larger than a certain threshold could result in a 100% reduction of the extra load-carrying capacity provided by FRP. The engineer can develop such correlations using a simple analytical method such as section analysis or more sophisticated means like finite element modeling.

### 7.4.2 AASHTO Bridge Load Rating Process

The AASHTO MBE offers guidance on bridge load rating and provides a standardized procedure for evaluation and a basis to determine the safe load capacity. It provides analytical methods such as Load and Resistance Factor Rating (LRFR), Allowable Stress Rating (ASR), and Load Factor Rating (LFR) methods, as well as empirical methods such as load ratings by load testing for determining the maximum

**TABLE 7.8**
**MBEI condition state [475]**

| Condition State | General Description |
|---|---|
| CS 1 | Good |
| CS 2 | Fair |
| CS 3 | Poor |
| CS 4 | Severe |

live load capacity of bridges. In a review of State DOT manuals for load rating by Phares et al. [479], it was found that LRFR has been predominantly used by most of the states, and the use of ASR or LFR is permissible but subject to the respective State DOT approval. It was also found that ASR is mostly used for timber and masonry structures. Thus, this section will focus on the LRFR approach of bridge load rating in the context of FRP/RSC application.

In AASHTO LRFR approach, the condition of a bridge element is assessed and given a condition state (Table 7.8) following the Manual for Bridge Element Inspection (MBEI) [475]. This condition state reflects the observed deterioration, damage, or degradation of the bridge elements due to factors like corrosion, fatigue, cracking, or spalling.

The MBE also provides a condition factor, φc, to adjust the capacity of a structural member based on its current condition state, as shown in Table 7.9. This condition factor is then applied to the nominal resistance or capacity of the element to account for the reduced effectiveness or strength due to its condition. Thus, load rating by the LRFR approach is typically an analytical method based on information taken from bridge plans/design, supplemented by information gathered from field inspections.

When condition rating is performed according to the SNBI condition rating scale discussed earlier in Section 7.2, an approximate conversion can be utilized to determine $\varphi_c$, as shown in Table 7.10.

This concept of load rating could also be implemented for FRP/RSC applications through a similar approach of condition factor developed for FRP materials or by incorporating FRP deterioration through reduction in their section properties based on engineering judgment.

**TABLE 7.9**
**AASHTO LRFR condition factor [480]**

| Condition State of Member | Condition Factor ($\varphi_c$) |
|---|---|
| Good or Satisfactory | 1.00 |
| Fair | 0.95 |
| Poor | 0.85 |

**TABLE 7.10**
**SNBI to MBEI condition scale conversion for selecting $\varphi_c$ [480]**

| SNBI Condition Rating | Equivalent AASHTO MBEI Member Condition State |
|---|---|
| 6 or higher | Good or Satisfactory |
| 5 | Fair |
| 4 or lower | Poor |

Alternatively, or until such general correspondence between the member condition state (or condition rating) and condition factors is developed, a general approach as follows can be adopted. If no major defect is reported or the condition is satisfactory (6) or better, the capacity of the elements will be at or near their intact condition for FRP-RC or as that calculated for strengthened elements for FRP-SC. In the case of reported defects requiring further analysis (e.g., Condition 5 and lower), as a general guideline for using the inspection results for determining the capacity for load rating purposes, the engineer in charge of load rating can follow the following overall approaches.

- For FRP-RC, determine the percentage of defect, e.g., cross-section loss, in the FRP bars reported by the inspection and relate the defect to the reduction of the strength in FRP. Conduct a section analysis that accounts for the level of deterioration of FRP strength in the capacity of the element, similar to the process for conventional steel-reinforced concrete.
- For FRP-SC, determine whether the defect reported (e.g., debonding, cracks) would affect the contribution of the FRP to the load-carrying capacity of the RC element and in what way (e.g., reducing composite action due to debonding, reduction of FRP strength because of cracking, etc.). Conduct a section analysis that incorporates the reduced contribution of the external FRP to the capacity of the element.
- For stand-alone elements, conduct section analysis to include the effect of reported damages.

### 7.4.3 Damage Thresholds and Remaining Life Prediction

In bridge engineering and in particular for bridges with FRP/RSC elements, there is limited research related to damage estimation, damage accumulation, the effect of damages on the load-carrying capacity, and the prediction of their remaining life. Some research has been conducted on those FRP elements that have been in service for a while, such as FRP bridge decks. This includes studies evaluating the post-damage responses of FRP decks by Lenett et al. [483]. The research explored the effects of damage on the performance of FRP sandwich deck panels and found that damage, particularly delamination, significantly alters deck stiffness and leads to abnormal deck behavior. However, research into long-term performance and damage accumulation has mainly focused on the durability testing of FRP material [484, 485], and fatigue testing of FRP decks [486]. Dutta et al. [486] conducted fatigue tests on various FRP and non-FRP decks under severe temperature conditions to understand the long-term effects of fatigue and temperature on FRP decks, and concluded that both factors negatively impact the decks' performance. Further research into the fatigue life of composite beams and civil components utilizing FRP has been conducted by Senne [487], Tang et al. [488], Eljufout [489], and Deng et al. [490]. However, these studies offer limited insights into the post-damage strength and behavior of bridge elements with FRP application, the impact of damage on their serviceability, and their remaining life.

This research gap can be addressed by building upon the extensive research being conducted in the defense and aerospace industries on the damage susceptibility, the immediate consequences of such damage, and the subsequent behavior of FRP materials and components. *The Military Handbook* 17, Volume 3 [482], offers a comprehensive examination of damage initiation, quantification, its impact on strength, and the methodologies the Department of Defense employs to evaluate damage and its influence on the functionality of FRP components in military contexts. In addition, there are several other publications related to the initiation of damage and the behavior following damage in FRP materials [491–496]. Kan [491] conducted a series of experimental investigations on different composite panels to assess how impact damage affects their strength and the remaining strength post-impact. McGovan and Ambur [492] explored the effects of impact damage on the residual strength of composite sandwich panels, both with and without compression, highlighting the relationship between impact magnitude, impact type, the size of the damaged area, and the panels' residual strength post-impact. Nyman [493] reviews existing theories and methodologies for establishing damage thresholds in composites, drawing conclusions from both experimental and analytical research. This research is particularly focused on evaluating damage severity, tolerance, and the remaining strength of composites used in the Swedish JAS39 Gripen fighter jet's development. Tomblin et al. [494, 495] conducted a semi-empirical investigation into the impact damage and fatigue resistance of sandwich airframe structures, proposing methods to utilize experimental findings to formulate damage assessment criteria. Leong et al. [496] address the issue of Barely Visible Impact Damage (BVID) and its significant effect on the compressive strength of FRP laminates, detailing the progression of damage in composites, especially under low-velocity impacts, influenced by factors like laminate dimensions, thickness, lay-up configuration, and material characteristics.

Further, numerous research works have been conducted in the analytical evaluation and modeling of damage and its progression in FRP materials. Case et al. [497, 498] presented the challenges of developing life prediction models for FRP materials commonly utilized in aerospace, providing a thorough discussion on the subject. A review paper by Nelon et al. [499] highlights the predictive capabilities of machine learning (ML) in addressing the damage state of FRP composite materials, including detecting a damage mechanism, locating a mechanism, identifying the type of mechanism, assessing the extent of the damage mechanism, and estimating

**TABLE 7.11**
**Available design guides and codes focused on FRP composites**

| Application | Internal Application | External Application | Material Specifications |
|---|---|---|---|
| AASHTO | LRFD Bridge Design Guide Specifications for GFRP-Reinforced Concrete (2nd ed.) [111] | Guide Specifications for Design of Bonded FRP Systems for Repair and Strengthening of Concrete Bridge Elements (1st ed.) [112] | ASTM International D7205/D7205M-06(2016). D7914/D7914M. D7957/D7957M-22 |
| ACI | ACI 440.11-22. Code Requirements for Structural Concrete Reinforced with Glass FRP Bars (1st ed.) [114] | ACI PRC-440.2-23: Design and Construction of Externally Bonded Fiber-Reinforced Polymer (FRP) Systems for Strengthening Concrete Structures [113] | ACI SPEC-440.5-22. Construction with Glass Fiber-Reinforced Polymer Reinforcing Bars [504] |
| CSA | CSA S806-12 (R2017). Design and Construction of Building Structures with Fibre-reinforced Polymers [185] | CSA S806-12 (R2017). Design and Construction of Building Structures with Fibre-reinforced Polymers [185] | CSA S807-19. Specification for Fibre-reinforced Polymers [505] |
| JPCI | Recommendation for Design and Construction of Concrete Structures Using Fiber-Reinforced Polymer (FRP) | – | – |

the useful remaining life of the material or system. Additional significant studies in the field address analytical challenges related to damage modeling, propagation, and residual life estimation in FRP materials [500–503]. While much of this research focuses on the damage thresholds and residual strength of thin FRP laminates or sandwich structures in aircraft, the insights gained from defense and aerospace studies serve as a valuable foundation for applying these concepts to FRP stand-alone and RSC elements. Ortiz et al. [132] conducted a thorough review presenting the current state-of-the-art on the main environmental and mechanical conditions affecting the durability and serviceability of FRP-RSC structures, including the most likely sources and their effects on the physical/mechanical properties of internal/external applications of FRP composites.

In summary, there is still much to learn and investigate regarding the long-term performance of FRP, especially for its application to bridge engineering. Nevertheless, there is an abundance of codes and specifications that cover the application of FRP (see Table 7.11). These documents have included factors for the uncertainties related to environmental and long-term effects that will in time be refined as more information becomes available.

# References


1. G. Marsh, "50 years of reinforced plastic boats," *Reinf. Plast.*, vol. 50, no. 9, pp. 16–19, 2006, doi: 10.1016/S0034-3617(06)71125-0.
2. G. Lubin, *Handbook of Composites*. Springer Science & Business Media, 2013.
3. G. Shier, *Flexural Behaviour of Fibre Reinforced Polymer Strengthened Reinforced Concrete Beams at Elevated Temperatures*. Queen's University (Canada), 2013.
4. O. Gunes, "A fracture based approach to understanding debonding in Frp bonded structural members," PhD diss., Department of Civil and Environmental Engineering, Massachusetts Institute of Technology, Cambridge, MA, 2004.
5. ACI Committee 440, *Guide for the Design and Construction of Externally Bonded FRP Systems for Strengthening Concrete Structures (ACI 440.2R-17)*. Farmington Hills, MI: American Concrete Institute, 2017.
6. S. Rizkalla, T. Hassan, and N. Hassan, "Design recommendations for the use of FRP for reinforcement and strengthening of concrete structures," *Prog. Struct. Eng. Mater.*, vol. 5, no. 1, pp. 16–28, 2003, doi: 10.1002/pse.139.
7. S. L. Iyer and R. G. Lampo, "Demonstration of advanced composite cables for prestressing applications in concrete waterfront structures," *US Army Corps Eng.*, 1998.
8. P. Zia, S. Ahmad, and M. Leming, *High-Performance Concretes, a State-of-Art Report (1989–1994)*. Report (1989–1994). Report No. FHWA-RD-97-030. McLean, VA: Federal Highway Administration, 1997.
9. T. Keller, *Use of Fibre Reinforced Polymers in Bridge Construction*. Vol. 7, Structural Engineering Documents. Zurich, Switzerland: International Association for Bridge and Structural Engineering, 2003.
10. A. Nanni, A. De Luca, and H. Jawaheri Zadeh, *Reinforced Concrete with FRP Bars: Mechanics and Design*. CRC Press, 2014.
11. L. C. Hollaway and J.-G. Teng, *Strengthening and Rehabilitation of Civil Infrastructures using Fibre-Reinforced Polymer (FRP) Composites*. Woodhead Publishing and CRC Press, 2008.
12. G. Garney, "Defects found through non-destructive testing methods of fiber reinforced polymeric composites," M.S. thesis, California State University, Fullerton, Fullerton, CA; Ann Arbor, MI: ProQuest, 2006.
13. Y. Gowayed, "Types of fiber and fiber arrangement in fiber-reinforced polymer (FRP) composites," in *Developments in Fiber-Reinforced Polymer (FRP) Composites for Civil Engineering*, edited by N. Uddin, pp. 3–17. Cambridge: Woodhead Publishing Limited, 2013.
14. J. W. S. Hearle, *High-Performance Fibres*. CRC Press, 2001.
15. D. B. Miracle and S. L. Donaldson, *ASM Handbook (Volume 21): Composites*. OH: ASM international, 2001.
16. B. Benmokrane, F. Elgabbas, E. A. Ahmed, and P. Cousin, "Characterization and comparative durability study of glass/vinylester, basalt/vinylester, and basalt/epoxy FRP bars," *J. Compos. Constr.*, vol. 19, no. 6, 2015, doi: 10.1061/(asce)cc.1943-5614.0000564.
17. F. Elgabbas, E. A. Ahmed, and B. Benmokrane, "Physical and mechanical characteristics of new basalt-FRP bars for reinforcing concrete structures," *Constr. Build. Mater.*, vol. 95, pp. 623–635, 2015, doi: 10.1016/j.conbuildmat.2015.07.036.

18. Z. Wu, X. Wang, and G. Wu, "Basalt FRP composite as reinforcements in infrastructure," *Seventeenth Annu. Int. Conf. Compos. Eng.*, pp. 21–24, 2009, [Online]. Available: http://wjoe.hebeu.edu.cn/ICCE-17 proceedings Hawaii USA/Wu, Zhishen (Ibaraki U., Hitachi, Japan) 1135.pdf.
19. X. Wang, G. Wu, and Z. Wu, "Tensile property of prestressing basalt FRP and hybrid FRP tendons under salt solution," *Proc. 6th Int. Conf. FRP Compos. Civ. Eng. CICE 2012*, pp. 1–8, 2012.
20. V. Lopresto, C. Leone, and I. De Iorio, "Mechanical characterisation of basalt fibre reinforced plastic," *Compos. Part B Eng.*, vol. 42, no. 4, pp. 717–723, 2011, doi: 10.1016/j.compositesb.2011.01.030.
21. S. K. Bhudolia, P. Perrotey, and S. C. Joshi, "Enhanced vibration damping and dynamic mechanical characteristics of composites with novel pseudo-thermoset matrix system," *Compos. Struct.*, vol. 179, pp. 502–513, 2017, doi: 10.1016/j.compstruct.2017.07.093.
22. S. K. Bhudolia, G. Gohel, L. Kah Fai, and R. J. Barsotti, "Fatigue response of ultrasonically welded carbon/Elium® thermoplastic composites," *Mater. Lett.*, vol. 264, p. 127362, 2020, doi: 10.1016/j.matlet.2020.127362.
23. S. K. Bhudolia, P. Perrotey, and S. C. Joshi, "Optimizing polymer infusion process for thin ply textile composites with novel matrix system," *Materials (Basel).*, vol. 10, no. 3, 2017, doi: 10.3390/ma10030293.
24. S. K. Bhudolia, P. Perrotey, and S. C. Joshi, "Mode I fracture toughness and fractographic investigation of carbon fibre composites with liquid Methylmethacrylate thermoplastic matrix," *Compos. Part B Eng.*, vol. 134, pp. 246–253, 2018, doi: 10.1016/j.compositesb.2017.09.057.
25. N. M. Telang, C. Dumlao, A. B. Mehrabi, A. T. Ciolko, and J. Gutierrez, *Field Inspection of In-Service FRP Bridge Decks.* NCHRP Report 564. Washington, DC: Transportation Research Board of the National Academies, 2006.
26. Z. Wu, Y. Wu, and M. F. M. Fahmy, *Structures Strengthened with Bonded Composites.* Woodhead Publishing An Imprint of Elsevier, 2020.
27. M. Chajes, T. Rollins, H. Dai, and T. Murphy, *Report on Techniques for Bridge Strengthening: Main Report.* Report No. FHWA-HIF-18-041. Washington, DC: U.S. Department of Transportation, Federal Highway Administration, Office of Infrastructure, 2019.
28. N. M. Telang, C. Dumlao, A. B. Mehrabi, A. T. Ciolko, and J. Gutierrez, "NCHRP report 564: Field inspection of in-service FRP bridge decks," *Transp. Res. Board*, 2006, doi: 10.17226/23284.
29. B. Andrawes, I. D. Shaw, and H. Zhao, "Repair & strengthening of distressed/amaged ends of prestressed beams with FRP composites," *Rep. No. FWHA-ICT-18-001 2. Illinois Cent. Transp. Urbana,* 2018.
30. G. Portnov, C. E. Bakis, E. Lackey, and V. Kulakov, "FRP reinforcing bars — Designs and methods of manufacture (Review of patents)," *Mech. Compos. Mater.*, vol. 49, no. 4, pp. 381–400, 2013.
31. A. ElSafty, B. Benmokrane, S. Rizkalla, H. Hassan, and M. Hassan, "Degradation assessment of internal continuous fiber reinforcement in concrete environment," *Univ. North Florida*, no. July, p. 270, 2014.
32. S. Biswas and J. Anurag, "Fabrication of composite laminates," *Reinf. Polym. Compos. Process. Charact. Post Life Cycle Assess.*, pp. 39–53, 2020, doi: 10.1002/9783527820979.ch3.
33. A. R. Hutchinson, *Surface Preparation of Component Materials.* Woodhead Publishing Limited, 2008.
34. A. Hutchinson, "Adhesives for externally bonded FRP reinforcement," *ICE Man. Constr. Mater.*, pp. 667–674, 2009.

35. L. C. Hollaway, *Applications of Advanced Fibre-Reinforced Polymer (FRP) Composites in Bridge Engineering: Rehabilitation of Metallic Bridge Structures, All-FRP Composite Bridges, and Bridges Built with Hybrid Systems*, no. 1996. Woodhead Publishing Limited, 2013.
36. T. Alkhrdaji, A. Nanni, G. Chen, and M. G. Barker, "Destructive and non-destructive testing of bridge J857, Phelps County, Missouri. Volume I-Strengthening and testing to failure of bridge decks," 2001.
37. Y. J. Kim, *Use of Fiber-Reinforced Polymers in Highway Infrastructure*. NCHRP Synthesis 512. Washington, DC: Transportation Research Board, 2017.
38. S. Moy, "Advanced fiber-reinforced polymer (FRP) composites for civil engineering applications," in *Developments in Fiber-Reinforced Polymer (FRP) Composites for Civil Engineering*, edited by N. Uddin, pp. 177–204. Cambridge, UK: Woodhead Publishing Limited, 2013.
39. J. G. Teng, J.-F. Chen, S. T. Smith, and L. Lam, *FRP: Strengthened RC Structures*. Chichester, West Sussex: John Wiley & Sons, 2002.
40. A. Zaman, S. A. Gutub, and M. A. Wafa, "A review on FRP composites applications and durability concerns in the construction sector," *J. Reinf. Plast. Compos.*, vol. 32, no. 24, pp. 1966–1988, 2013, doi: 10.1177/0731684413492868.
41. L. C. Hollaway, "A review of the present and future utilisation of FRP composites in the civil infrastructure with reference to their important in-service properties," *Constr. Build. Mater.*, vol. 24, no. 12, pp. 2419–2445, 2010.
42. V. Benzecry, J. Brown, A. Al-Khafaji, R. Haluza, R. Koch, M. Nagarajan, C. E. Bakis, J. J. Myers, and A. Nanni, *Durability of GFRP Bars Extracted from Bridges with 15 to 20 Years of Service Life*. Farmington Hills, MI: ACI Foundation, 2019.
43. F. Micelli and A. Nanni, "Durability of FRP rods for concrete structures," *Constr. Build. Mater.*, vol. 18, no. 7, pp. 491–503, 2004, doi: 10.1016/j.conbuildmat.2004.04.012.
44. Y. Chen, J. F. Davalos, I. Ray, and H. Y. Kim, "Accelerated aging tests for evaluations of durability performance of FRP reinforcing bars for concrete structures," *Compos. Struct.*, vol. 78, no. 1, pp. 101–111, 2007, doi: 10.1016/j.compstruct.2005.08.015.
45. B. Benmokrane, P. Wang, T. M. Ton-That, H. Rahman, and J.-F. Robert, "Durability of glass fiber-reinforced polymer reinforcing bars in concrete environment," *J. Compos. Constr.*, vol. 6, no. 3, pp. 143–153, 2002.
46. F. Ceroni, E. Cosenza, M. Gaetano, and M. Pecce, "Durability issues of FRP rebars in reinforced concrete members," *Cem. Concr. Compos.*, vol. 28, no. 10, pp. 857–868, 2006.
47. Y. J. Kim, "State of the practice of FRP composites in highway bridges," *Eng. Struct.*, vol. 179, pp. 1–8, 2019.
48. D. A. Hensher, *Fiber-Reinforced-Plastic (FRP) Reinforcement for Concrete Structures: Properties and Applications*, vol. 42. Elsevier, 2016.
49. W. P. Wu, "Thermomechanical properties of fiber-reinforced plastic (FRP) bars." West Virginia University, 1991.
50. A. Nanni, S. Rizkalla, C. E. Bakis, J. O. Conrad, and A. A. Abdelrahman, "Characterization of GFRP ribbed rod used for reinforced concrete construction," *Proc. Int. Compos. Exhib.*, pp. 11–16, 1998.
51. P. K. Mallick, *Fiber-Reinforced Composites: Materials, Manufacturing, and Design*. CRC Press, 2007.
52. A. Nanni, A. De Luca, and H. Jawaheri Zadeh, *Reinforced Concrete with FRP Bars: Mechanics and Design*. CRC Press, 2019.
53. A. Mirmiran, W. Yuan, and X. Chen, "Design for slenderness in concrete columns internally reinforced with fiber-reinforced polymer bars," *Struct. J.*, vol. 98, no. 1, pp. 116–125, 2001.

54. Florida Department of Transportation. *Structures Manual.* Tallahassee, FL: Florida Department of Transportation, 2025.
55. M. Rossini and A. Nanni, *MILDGLASS: GFRP Strand for Resilient Mild Pre-Stressed Concrete.* Final Report for NCHRP IDEA Project 207. Washington, DC: Transportation Research Board, 2020.
56. V. Benzecry, M. Rossini, C. Morales, S. Nolan, and A. Nanni, "Design of marine dock using concrete mixed with seawater and FRP bars," *J. Compos. Constr.*, vol. 25, no. 1, p. 5020006, 2021, doi: 10.1061/(ASCE)CC.1943-5614.0001100.
57. M. Ekenel, F. D. C. y Basalo and A. Nanni, "Fiber-reinforced polymer reinforcement for concrete members," *Concr. Int.*, vol. 43, no. 2, pp. 18–22, 2021.
58. K. A. Soudki, "FRP reinforcement for prestressed concrete structures," *Prog. Struct. Eng. Mater.*, vol. 1, no. 2, pp. 135–142, 1998.
59. ACI Committee 440, "State-of-the-art report on fiber reinforced plastic (FRP) reinforcement for concrete structures," 1996.
60. N. F. Grace, F. C. Navarre, R. B. Nacey, W. Bonus, and L. Collavino, "Design-construction of bridge street bridge-first CFRP bridge in the United States," *PCI J.*, vol. 47, no. 5, pp. 20–35, 2002.
61. S. C. Das and M. Nizam, "Applications of fiber reinforced polymer composites (FRP) in civil engineering," *Int. J. Adv. Struct. Geotech. Eng.*, vol. 3, no. 3, pp. 299–309, 2014.
62. L. C. Hollaway, *Advanced Polymer Composites and Polymers in the Civil Infrastructure.* Elsevier, 2001.
63. M. Motavalli and C. Czaderski, "FRP composites for retrofitting of existing civil structures in Europe: State-of-the-art review," in *International Conference of Composites & Polycon*, pp. 17–19. Arlington, VA: American Composites Manufacturers Association, 2007.
64. T. Uomoto, H. Mutsuyoshi, F. Katsuki, and S. Misra, "Use of fiber reinforced polymer composites as reinforcing material for concrete," *J. Mater. Civ. Eng.*, vol. 14, no. 3, pp. 191–209, 2002.
65. B. M. Carmichael and R. W. Barnes, "Repair of the Uphapee Creek Bridge with FRP laminates," no. November, p. 120, 2005.
66. P. A. Ritchie, D. A. Thomas, L.-W. Lu, and G. M. Connelly, "External reinforcement of concrete beams using fiber reinforced plastic," Lehigh University, 1988.
67. M. Arduini and A. Nanni, "Parametric study of beams with externally bonded FRP reinforcement," *ACI Struct. J.*, vol. 94, no. 5, pp. 493–501, 1997.
68. H. Toutanji, L. Zhao, and Y. Zhang, "Flexural behavior of reinforced concrete beams externally strengthened with CFRP sheets bonded with an inorganic matrix," *Eng. Struct.*, vol. 28, no. 4, pp. 557–566, 2006.
69. M. Z. Naser, R. A. Hawileh, and J. A. Abdalla, "Fiber-reinforced polymer composites in strengthening reinforced concrete structures: A critical review," *Eng. Struct.*, vol. 198, p. 109542, 2019.
70. M. J. Chajes, T. F. Januszka, D. R. Mertz, T. A. Thomson, and W. W. Finch, "Shear strengthening of reinforced concrete beams using externally applied composite fabrics," *Struct. J.*, vol. 92, no. 3, pp. 295–303, 1995.
71. T. El-Maaddawy and Y. Chekfeh, "Retrofitting of severely shear-damaged concrete t-beams using externally bonded composites and mechanical end anchorage," *J. Compos. Constr.*, vol. 16, no. 6, pp. 693–704, 2012.
72. A. Carolin and B. Täljsten, "Experimental study of strengthening for increased shear bearing capacity," *J. Compos. Constr.*, vol. 9, no. 6, pp. 488–496, 2005.
73. H. A. Rasheed, *Strengthening Design of Reinforced Concrete with FRP.* CRC Press, 2014.
74. S. T. Smith and J. G. Teng, "FRP-strengthened RC beams. II: Assessment of debonding strength models," *Eng. Struct.*, vol. 24, no. 4, pp. 397–417, 2002.

75. S. T. Smith and J. G. Teng, "FRP-strengthened RC beams. I: review of debonding strength models," *Eng. Struct.*, vol. 24, no. 4, pp. 385–395, 2002.
76. T. Yamamoto, "FRP strengthening of RC columns for seismic retrofitting," in *Proc. 10th World Conf. on Earthquake Engineering, Madrid, Balkema*, 1992, pp. 5205–5210.
77. K. K. Antoniades, T. N. Salonikios, and A. J. Kappos, "Tests on seismically damaged reinforced concrete walls repaired and strengthened using fiber-reinforced polymers," *J. Compos. Constr.*, vol. 9, no. 3, pp. 236–246, 2005.
78. A. S. Mosallam and A. Nasr, "Structural performance of RC shear walls with post-construction openings strengthened with FRP composite laminates," *Compos. Part B Eng.*, vol. 115, pp. 488–504, 2017.
79. P. R. Salom, J. Gergely, and D. T. Young, "Torsional strengthening of spandrel beams with fiber-reinforced polymer laminates," *J. Compos. Constr.*, vol. 8, no. 2, pp. 157–162, 2004.
80. K. I. Mohammad and D. B. Jaafar Al-Sulayfani, "Torsional strengthening of RC beams with CFRP wrap," *Tikrit J. Eng. Sci.*, vol. 20, no. 3, pp. 1–9, 2013.
81. M. R. Mohammadizadeh and M. J. Fadaei, "Torsional behavior of high-strength concrete beams strengthened using CFRP sheets; an experimental and analytical study," 2009.
82. L. P. Ye, K. Zhang, S. H. Zhao, and P. Feng, "Experimental study on seismic strengthening of RC columns with wrapped CFRP sheets," *Constr. Build. Mater.*, vol. 17, no. 6–7, pp. 499–506, 2003.
83. R. A. Barnes and G. C. Mays, "Fatigue performance of concrete beams strengthened with CFRP plates," *J. Compos. Constr.*, vol. 3, no. 2, pp. 63–72, 1999.
84. C. G. Papakonstantinou, M. F. Petrou, and K. A. Harries, "Fatigue behavior of RC beams strengthened with GFRP sheets," *J. Compos. Constr.*, vol. 5, no. 4, pp. 246–253, 2001.
85. H. M. Tanarslan and S. Altin, "Behavior of RC T-section beams strengthened with CFRP strips, subjected to cyclic load," *Mater. Struct.*, vol. 43, no. 4, pp. 529–542, 2010.
86. H. Toutanji, L. Zhao, Y. Deng, Y. Zhang, and P. Balaguru, "Cyclic behavior of RC beams strengthened with carbon fiber sheets bonded by inorganic matrix," *J. Mater. Civ. Eng.*, vol. 18, no. 1, pp. 28–35, 2006.
87. A. Hosny, H. Shaheen, A. Abdelrahman, and T. Elafandy, "Performance of reinforced concrete beams strengthened by hybrid FRP laminates," *Cem. Concr. Compos.*, vol. 28, no. 10, pp. 906–913, 2006.
88. S. O. Asplund, "Strengthening bridge slabs with grouted reinforcement," *Journal Proceedings*, vol. 45, no. 1, pp. 397–406, 1949.
89. G. E. Warren, "Waterfront repair and upgrade, advanced technology demonstration site No. 3: NAVSTA Bravo 25, Pearl Harbour," *Site Specif. Rep. SSR-2567-SHR, Nav. Facil. Eng. Serv. Center, Port Hueneme, CA*, 2000.
90. S. W. Garrity, "Near-surface reinforcement of masonry arch highway bridges," 2001.
91. L. De Lorenzis, A. Nanni, and A. La Tegola, "Strengthening of reinforced concrete structures with near surface mounted FRP rods," in *International Meeting on Composite Materials, PLAST*, 2000, pp. 9–11.
92. A. Astorga, H. Santa Maria, and M. Lopez, "Behavior of a concrete bridge cantilevered slab reinforced using NSM CFRP strips," *Constr. Build. Mater.*, vol. 40, pp. 461–472, 2013.
93. J. A. O. Barros, S. J. E. Dias, and J. L. T. Lima, "Efficacy of CFRP-based techniques for the flexural and shear strengthening of concrete beams," *Cem. Concr. Compos.*, vol. 29, no. 3, pp. 203–217, 2007.
94. F. Al-Mahmoud, A. Castel, R. François, and C. Tourneur, "Strengthening of RC members with near-surface mounted CFRP rods," *Compos. Struct.*, vol. 91, no. 2, pp. 138–147, 2009.

95. S. M. Soliman, E. El-Salakawy, and B. Benmokrane, "Flexural behaviour of concrete beams strengthened with near surface mounted fibre reinforced polymer bars," *Can. J. Civ. Eng.*, vol. 37, no. 10, pp. 1371–1382, 2010.
96. A. Hajihashemi, D. Mostofinejad, and M. Azhari, "Investigation of RC beams strengthened with prestressed NSM CFRP laminates," *J. Compos. Constr.*, vol. 15, no. 6, pp. 887–895, 2011.
97. S. Hong and S.-K. Park, "Effect of prestress and transverse grooves on reinforced concrete beams prestressed with near-surface-mounted carbon fiber-reinforced polymer plates," *Compos. Part B Eng.*, vol. 91, pp. 640–650, 2016.
98. F. Oudah and R. El-Hacha, "Performance of RC beams strengthened using prestressed NSM-CFRP strips subjected to fatigue loading," *J. Compos. Constr.*, vol. 16, no. 3, pp. 300–307, 2012.
99. M. R. M. Hosseini, S. J. E. Dias, and J. A. O. Barros, "Effectiveness of prestressed NSM CFRP laminates for the flexural strengthening of RC slabs," *Compos. Struct.*, vol. 111, pp. 249–258, 2014.
100. M. R. M. Hosseini, S. J. E. Dias, and J. A. O. Barros, "Flexural strengthening of reinforced low strength concrete slabs using prestressed NSM CFRP laminates," *Compos. Part B Eng.*, vol. 90, pp. 14–29, 2016.
101. J. A. O. Barros and S. J. E. Dias, "Near surface mounted CFRP laminates for shear strengthening of concrete beams," *Cem. Concr. Compos.*, vol. 28, no. 3, pp. 276–292, 2006.
102. A. Rizzo and L. De Lorenzis, "Behavior and capacity of RC beams strengthened in shear with NSM FRP reinforcement," *Constr. Build. Mater.*, vol. 23, no. 4, pp. 1555–1567, 2009.
103. J. A. O. Barros and S. J. E. Dias, "Assessment of the effectiveness of the NSM shear strengthening technique for deep T cross section RC beams," 2013.
104. M. Jalali, M. K. Sharbatdar, J.-F. Chen, and F. J. Alaee, "Shear strengthening of RC beams using innovative manually made NSM FRP bars," *Constr. Build. Mater.*, vol. 36, pp. 990–1000, 2012.
105. L. De Lorenzis and J.-G. Teng, "Near-surface mounted FRP reinforcement: An emerging technique for strengthening structures," *Compos. Part B Eng.*, vol. 38, no. 2, pp. 119–143, 2007.
106. S. S. K. Dolati and A. Mehrabi, "FRP sheet/jacket system as an alternative method for splicing prestressed-precast concrete piles," *Case Stud. Constr. Mater.*, p. e00912, 2022.
107. S. S. Khedmatgozar Dolati and A. Mehrabi, "FRP splice system for joining structural elements." US Patent 11,319,706, May 2022.
108. A. Mehrabi and S. S. Khedmatgozar Dolati, "NSMB pile splice system for precast concrete piles." US Patent 11,319,689, May 2022.
109. A. B. Mehrabi and S. S. Khedmatgozar Dolati, "Alternative materials and configurations for prestressed-precast concrete pile splice connection," *ABC-UTC Rep.*, 2021.
110. JPCI, *Recommendation for Design and Construction of Concrete Structures using Fiber Reinforced Polymer*. Japan Prestressed Concrete Institute, Tokyo, Japan, 2021.
111. AASHTO, *LRFD Bridge Design Guide Specifications for GFRP-Reinforced Concrete*, 2nd ed. American Association of State Highway and Transportation Officials, Washington, DC, 2018.
112. AASHTO, *Guide Specifications for Design of Bonded FRP Systems for Repair and Strengthening of Concrete Bridge Elements*, 1st ed. American Association of State Highway and Transportation Officials, Washington, DC, 2012.
113. ACI Committee 440, "ACI PRC-440.2-23: Design and construction of externally bonded Fiber-Reinforced Polymer (FRP) systems for strengthening concrete structures—Guide," 2023.

114. ACI Committee 440, *Code Requirements for Structural Concrete Reinforced with Glass FRP Bars (ACI 440.11-22)*. American Concrete Institute, Farmington Hills, MI, 2022.
115. ACI 440.1R-15, *Guide for the Design and Construction of Structural Concrete Reinforced with Fiber-Reinforced Polymer (FRP) Bars (ACI 440.1R-15)*. American Concrete Institute, Farmington Hills, MI, 2015.
116. T. Uomoto, "Durability design of GFRP rods for concrete reinforcement," *Fibre-Reinforced Polym. Reinf. Concr. Struct. (In 2 Vol.)*, pp. 37–50, 2003.
117. S. F. Breña, R. M. Bramblett, M. A. Benouaich, S. L. Wood, and M. E. Kreger, *Use of Carbon Fiber Reinforced Polymer Composites to Increase the Flexural Capacity of Reinforced Concrete Beams. Research Report 1776-1*, vol. 7, no. 2. Center for Transportation Research. The University of Texas at Austin, 2001.
118. ACI Committee 201, *Guide to Durable Concrete (ACI 201.2R-16)*. American Concrete Institute, Farmington Hills, MI, 2016.
119. T. Q. Liu, X. Liu, and P. Feng, "A comprehensive review on mechanical properties of pultruded FRP composites subjected to long-term environmental effects," *Compos. Part B Eng.*, vol. 191, no. January, p. 107958, 2020, doi: 10.1016/j.compositesb.2020.107958.
120. Y. Duo, X. Liu, Y. Liu, T. Tafsirojjaman, and M. Sabbrojjaman, "Environmental impact on the durability of FRP reinforcing bars," *J. Build. Eng.*, vol. 43, no. February, p. 102909, 2021, doi: 10.1016/j.jobe.2021.102909.
121. C. Lu, M. Ni, T. Chu, and L. He, "Comparative investigation on tensile performance of FRP bars after exposure to water, seawater, and alkaline solutions," *J. Mater. Civ. Eng.*, vol. 32, no. 7, p. 04020170, 2020, doi: 10.1061/(asce)mt.1943-5533.0003243.
122. J. P. Won, S. J. Lee, Y. J. Kim, C. Il Jang, and S. W. Lee, "The effect of exposure to alkaline solution and water on the strength-porosity relationship of GFRP rebar," *Compos. Part B Eng.*, vol. 39, no. 5, pp. 764–772, 2008, doi: 10.1016/j.compositesb.2007.11.002.
123. H. Y. Kim, Y. H. Park, Y. J. You, and C. K. Moon, "Short-term durability test for GFRP rods under various environmental conditions," *Compos. Struct.*, vol. 83, no. 1, pp. 37–47, 2008, doi: 10.1016/j.compstruct.2007.03.005.
124. J. S. Earl and R. A. Shenoi, "Hygrothermal ageing effects on FRP laminate and structural foam materials," *Compos. Part A Appl. Sci. Manuf.*, vol. 35, no. 11, pp. 1237–1247, 2004, doi: 10.1016/j.compositesa.2004.04.007.
125. I. B. C. M. Rocha, S. Raijmaekers, R. P. L. Nijssen, F. P. van der Meer, and L. J. Sluys, "Hygrothermal ageing behaviour of a glass/epoxy composite used in wind turbine blades," *Compos. Struct.*, vol. 174, pp. 110–122, 2017, doi: 10.1016/j.compstruct.2017.04.028.
126. S. Alessi, G. Pitarresi, and G. Spadaro, "Effect of hydrothermal ageing on the thermal and delamination fracture behaviour of CFRP composites," *Compos. Part B Eng.*, vol. 67, pp. 145–153, 2014, doi: 10.1016/j.compositesb.2014.06.006.
127. M. Wang, X. Xu, J. Ji, Y. Yang, J. Shen, and M. Ye, "The hygrothermal aging process and mechanism of the novolac epoxy resin," *Compos. Part B Eng.*, vol. 107, no. C, pp. 1–8, 2016, doi: 10.1016/j.compositesb.2016.09.067.
128. E. A. Byars, P. Waldron, V. Dejke, S. Demis, and S. Heddadin, "Durability of FRP in concrete – Deterioration mechanisms," *Int. J. Mater. Prod. Technol.*, vol. 19, no. 1–2, pp. 28–39, 2003, doi: 10.1504/IJMPT.2003.003554.
129. M. P. Lebedev, O. V. Startsev, and A. K. Kychkin, "The effects of aggressive environments on the mechanical properties of basalt plastics," *Heliyon*, vol. 6, no. 3, p. e03481, 2020, doi: 10.1016/j.heliyon.2020.e03481.
130. Y. A. Al-Salloum, S. El-Gamal, T. H. Almusallam, S. H. Alsayed, and M. Aqel, "Effect of harsh environmental conditions on the tensile properties of GFRP bars," *Compos. Part B Eng.*, vol. 45, no. 1, pp. 835–844, 2013, doi: 10.1016/j.compositesb.2012.05.004.

131. A. Serbescu, M. Guadagnini, and K. Pilakoutas, "Mechanical characterization of Basalt FRP rebars and long-rerm strength predictive model," *J. Compos. Constr.*, vol. 19, no. 2, p. 04014037, 2015, doi: 10.1061/(asce)cc.1943-5614.0000497.
132. J. D. Ortiz, S. S. Khedmatgozar Dolati, P. Malla, A. Nanni, and A. Mehrabi, "FRP-reinforced/strengthened concrete: State-of-the-art review on durability and mechanical effects," *Materials (Basel).*, vol. 16, no. 5, pp. 1–30, 2023.
133. F. E. Tannous and H. Saadatmanesh, "Durability of AR glass fiber reinforced plastic bars," *J. Compos. Constr.*, vol. 3, no. February, pp. 12–19, 1999.
134. H. Saadatmanesh and F. E. Tannous, "Long-term behavior of Aramid Fiber-Reinforced Plastic (AFRP) tendons," *ACI Mater. J.*, vol. 96, no. 3, pp. 297–305, 1999.
135. X. Zhang and J. Ou, "Durability experimental research on resistance of acidic, alkali, salt solutions and freeze-thaw properties of FRP bar," *J. Wuhan Univ. Technol.*, vol. 29, pp. 33–36+54, Jan. 2007.
136. Y. J. Kim and P. J. Heffernan, "Fatigue behavior of externally strengthened concrete beams with fiber-reinforced polymers: State of the art," *J. Compos. Constr.*, vol. 12, no. 3, pp. 246–256, 2008, doi: 10.1061/(asce)1090-0268(2008)12:3(246).
137. G. Wu, X. Wang, Z. Wu, Z. Dong, and Q. Xie, "Degradation of basalt FRP bars in alkaline environment," *Sci. Eng. Compos. Mater.*, vol. 22, no. 6, pp. 649–657, 2015, doi: 10.1515/secm-2014-0040.
138. G. Wu, Z.-Q. Dong, X. Wang, Y. Zhu, and Z.-S. Wu, "Prediction of long-term performance and durability of BFRP bars under the combined effect of sustained load and corrosive solutions," *J. Compos. Constr.*, vol. 19, no. 3, p. 04014058, 2015, doi: 10.1061/(asce)cc.1943-5614.0000517.
139. M. Frigione and M. Lettieri, "Durability issues and challenges for material advancements in FRP employed in the construction industry," *Polymers (Basel).*, vol. 10, no. 3, 2018, doi: 10.3390/polym10030247.
140. R. Sen, "Developments in the durability of FRP-concrete bond," *Construction and Building Materials*, vol. 78. Elsevier Ltd, pp. 112–125, Mar. 01, 2015, doi: 10.1016/j.conbuildmat.2014.12.106.
141. M. A. Abanilla, V. M. Karbhari, and Y. Li, "Interlaminar and intralaminar durability characterization of wet layup carbon/epoxy used in external strengthening," *Compos. Part B Eng.*, vol. 37, no. 7–8, pp. 650–661, 2006, doi: 10.1016/j.compositesb.2006.02.023.
142. M. S. Sciolti, M. Frigione, and M. A. Aiello, "Wet lay-up manufactured FRPs for concrete and masonry repair: Influence of water on the properties of composites and on their epoxy components," *J. Compos. Constr.*, vol. 14, no. 6, pp. 823–833, 2010, doi: 10.1061/(asce)cc.1943-5614.0000132.
143. S. H. Alsayed, A. Alhozaimy, Y. A. Al-Salloum, and T. H. Almusallam, "Durability of the new generation of GFRP rebars under severe environments," in *Proceedings of the Second International Conference on Durability of Fiber Reinforced Polymer (FRP) Composites for Construction (CDCC 2002),* May 2002, pp. 651–663.
144. Portland Cement Association, "Alkali-aggregate reaction," Portland Cement Association, February 2022.
145. D. K. Panesar, *Supplementary Cementing Materials*. Elsevier LTD, 2019.
146. V. M. Karbhari and K. Ghosh, "Comparative durability evaluation of ambient temperature cured externally bonded CFRP and GFRP composite systems for repair of bridges," *Compos. Part A Appl. Sci. Manuf.*, vol. 40, no. 9, pp. 1353–1363, 2009, doi: 10.1016/j.compositesa.2009.01.011.
147. S. K. Rege and S. C. Lakkad, "Effect of salt water on mechanical properties of fibre reinforced plastics," *Fibre Sci. Technol.*, vol. 19, no. 4, pp. 317–324, 1983, doi: 10.1016/0015-0568(83)90017-9.

148. A. Belarbi, J. Myers, K. Chandarshekhara, S. Watkins, and S. Bae, *Structural Integrity of RC Columns Wrapped with FRP Sheet Subjected to Various Environmental Conditions Including Corrosion*. Research Report No. RDT 04-021. Jefferson City, MO: Missouri Department of Transportation, Research, Development and Technology, 2004.
149. A. H. Rahman, C. Y. Kingsley, and J. Crimi, "Durability of a FRP grid reinforcement," in *Proceedings of the Advanced Composite Materials in Bridges and Structures: 2nd International Conference*, pp. 681–690. Montréal, QC, Canada, 1996.
150. A. Singhvi and A. Mirmiran, "Creep and durability of environmentally conditioned FRP-RC beams using fiber optic sensors," *J. Reinf. Plast. Compos.*, vol. 21, no. 4, pp. 351–373, 2002, doi: 10.1106/073168402026254.
151. L. Bian, J. Xiao, J. Zeng, and S. Xing, "Effects of seawater immersion on water absorption and mechanical properties of GFRP composites," *J. Compos. Mater.*, vol. 46, no. 25, pp. 3151–3162, 2012, doi: 10.1177/0021998312436992.
152. Z. Wang *et al.*, "Long-term durability of basalt- and glass-fibre reinforced polymer (BFRP/GFRP) bars in seawater and sea sand concrete environment," *Constr. Build. Mater.*, vol. 139, pp. 467–489, 2017, doi: 10.1016/j.conbuildmat.2017.02.038.
153. C. N. Morales, G. Claure, A. R. Emparanza, and A. Nanni, "Durability of GFRP reinforcing bars in seawater concrete," *Constr. Build. Mater.*, vol. 270, p. 121492, 2021, doi: 10.1016/j.conbuildmat.2020.121492.
154. R. Sen, M. Shahawy, S. Sukumar, and J. Rosas, "Durability of carbon pretensioned elements in a marine environment," *ACI Struct. J.*, vol. 95, no. 6, pp. 716–724, 1998, doi: 10.14359/585.
155. F. Rostasy, "On durability of FRP in aggressive environments," in *Proceedings of the Third International Symposium on Non-Metallic (FRP) Reinforcement for Concrete Structures*, 1997, pp. 107–114.
156. M. A. G. Silva, M. T. Cidade, H. Biscaia, and R. Marreiros, "Composites and FRP-strengthened beams subjected to dry/wet and salt fog cycles," *J. Mater. Civ. Eng.*, vol. 26, no. 12, p. 4014092, 2014.
157. M. J. Chajes, W. W. Finch, T. F. Januszka, and T. A. Thomson, "Bond and force transfer of composite-material plates bonded to concrete," *ACI Struct. J.*, vol. 93, no. 2, 1995, doi: 10.14359/1491.
158. H. A. Toutanji and W. Gómez, "Durability characteristics of concrete beams externally bonded with FRP composite sheets," *Cem. Concr. Compos.*, vol. 19, no. 4, pp. 351–358, 1997, doi: 10.1016/S0958-9465(97)00028-0.
159. P. Cousin, M. Hassan, P. V. Vijay, M. Robert, and B. Benmokrane, "Chemical resistance of carbon, basalt, and glass fibers used in FRP reinforcing bars," *J. Compos. Mater.*, vol. 53, no. 26–27, pp. 3651–3670, 2019, doi: 10.1177/0021998319844306.
160. M. L. Porter and B. A. Barnes, "Accelerated aging degradation of glass fiber composites," in *Second International Conference on Composites in Infrastructure*, 1998, pp. 446–459.
161. H. Li, G. Xian, M. Ma, and J. Wu, "Durability and fatigue performances of basalt fiber / epoxy reinforcing bars," *Proc. 6th Int. Conf. FRP Compos. Civ. Eng. CICE 2012*, pp. 1–8, 2012.
162. B. Benmokrane, M. Hassan, M. Robert, P. V. Vijay, and A. Manalo, "Effect of different constituent fiber, resin, and sizing combinations on alkaline resistance of basalt, carbon, and glass FRP bars," *J. Compos. Constr.*, vol. 24, no. 3, p. 04020010, 2020, doi: 10.1061/(asce)cc.1943-5614.0001009.
163. B. Benmokrane, V. L. Brown, A. H. Ali, K. Mohamed, and C. Shield, "Reconsideration of the environmental reduction factor CE for GFRP reinforcing bars in concrete structures," *J. Compos. Constr.*, vol. 24, no. 4, pp. 1–6, 2020, doi: 10.1061/(asce)cc.1943-5614.0001040.

164. Y. Chen, J. F. Davalos, and I. Ray, "Durability prediction for GFRP reinforcing bars using short-term data of accelerated aging tests," *J. Compos. Constr.*, vol. 10, no. 4, pp. 279–286, 2006, doi: 10.1061/(asce)1090-0268(2006)10:4(279).
165. M. Robert, P. Cousin, and B. Benmokrane, "Durability of GFRP reinforcing bars embedded in moist concrete," *J. Compos. Constr.*, vol. 13, no. 2, pp. 66–73, 2009, doi: 10.1061/(asce)1090-0268(2009)13:2(66).
166. H. Fergani, M. Di Benedetti, C. Miàs Oller, C. Lynsdale, and M. Guadagnini, "Durability and degradation mechanisms of GFRP reinforcement subjected to severe environments and sustained stress," *Constr. Build. Mater.*, vol. 170, pp. 637–648, 2018, doi: 10.1016/j.conbuildmat.2018.03.092.
167. J. Xie, Y. Li, Z. Lu, Z. Fan, J. Li, and S. Li, "Effects of immersion in water, alkaline solution, and seawater on the shear performance of basalt FRP bars in seawater–sea sand concrete," *J. Compos. Constr.*, vol. 26, no. 2, pp. 1–14, 2022, doi: 10.1061/(asce)cc.1943-5614.0001184.
168. E. Gudonis, E. Timinskas, V. Gribniak, G. Kaklauskas, A. K. Arnautov, and V. Tamulėnas, "FRP reinforcement for concrete structures: State-of-the-art review of application and design," *Eng. Struct. Technol.*, vol. 5, no. 4, pp. 147–158, 2014, doi: 10.3846/2029882x.2014.889274.
169. M. Scheibe, F. S. Rostasy, and L. Taerwe, "Stress-rupture of AFRP subjected to alkaline solutions and elevated temperature – experiments, 2nd International symposium, Non-metallic \(FRP\) reinforcement for concrete structures," in *Non-metallic \(FRP\) Reinforcement for Concrete Structures, RILEM PROCEEDINGS, 2nd International Symposium, Non-metallic \(FRP\) Reinforcement for Concrete Structures*, 1995, no. 29, pp. 67–73.
170. M. Mohammadi, D. Mostofinejad, and M. Barghian, "Effects of surface preparation method on FRP-concrete bond strength under alkaline conditions," *J. Compos. Constr.*, vol. 21, no. 4, p. 04017010, 2017, doi: 10.1061/(asce)cc.1943-5614.0000786.
171. S. W. Tang, Y. Yao, C. Andrade, and Z. J. Li, "Recent durability studies on concrete structure," *Cem. Concr. Res.*, vol. 78, pp. 143–154, 2015, doi: 10.1016/j.cemconres.2015.05.021.
172. S. Demis and V. G. Papadakis, "Durability of glass Fibre Reinforced Polymer (FRP) materials in concrete – Effect of carbonation on bond strength," Apr. 2011.
173. J. Chin, T. Nguyen, and K. Aouadi, "Effects of environmental exposure on Fiber-Reinforced Plastic (FRP) materials used in construction," *J. Compos. Technol. Res.*, vol. 19, Oct. 1997, doi: 10.1520/CTR10120J.
174. S. Sethi and B. C. Ray, "Environmental effects on fibre reinforced polymeric composites: Evolving reasons and remarks on interfacial strength and stability," *Adv. Colloid Interface Sci.*, vol. 217, pp. 43–67, 2015, doi: 10.1016/j.cis.2014.12.005.
175. J. R. Cromwell, K. A. Harries, and B. M. Shahrooz, "Environmental durability of externally bonded FRP materials intended for repair of concrete structures," *Constr. Build. Mater.*, vol. 25, no. 5, pp. 2528–2539, 2011, doi: 10.1016/j.conbuildmat.2010.11.096.
176. V. M. Karbhari *et al.*, "Durability gap analysis for fiber-reinforced polymer composites in civil infrastructure," *J. Compos. Constr.*, vol. 7, no. 3, pp. 238–247, 2003, doi: 10.1061/(asce)1090-0268(2003)7:3(238).
177. G. L. Balázs and A. Borosnyói, "Long-term behavior of FRP," in *Composites in Construction: A Reality*, edited by E. Cosenza, G. Manfredi, and A. Nanni, pp. 84–91. Reston, VA: American Society of Civil Engineers, 2001.
178. L. C. Bank, *Composites for Construction: Structural Design with FRP Materials*. John Wiley & Sons, 2006.

179. I. Sasaki, I. Nishizaki, H. Sakamoto, K. Katawaki, and Y. Kawamoto, "Durability evaluation of FRP cables by exposure tests," in *Proceedings of the Third International Symposium on Non-Metallic (FRP) Reinforcement for Concrete Structures (FRPRCS-3)*, 1997, pp. 131–137.
180. T. Uomoto, "Durability of FRP as reinforcement for concrete structures," in *Proceedings of the 3rd International Conference on Advanced Composite Materials in Bridges and Structures, ACMBS-3*, 2000, pp. 3–17.
181. M. Wang and Y. Yang, "Study on aging properties and degradation mechanism of FRP bars," in *E3S Web of Conferences*, 2021, vol. 261, pp. 1–4, doi: 10.1051/e3sconf/202126102079.
182. L. C. Bank, T. R. Gentry, and A. Barkatt, "Accelerated test methods to determine the long-term behavior of FRP composite structures: Environmental effects," *J. Reinf. Plast. Compos.*, vol. 14, no. 6, pp. 559–587, 1995, doi: 10.1177/073168449501400602.
183. T. Odagiri, K. Matsumoto, and H. Nakai, "Fatigue and relaxation characteristics of continuous aramid fiber reinforced plastic rods," in *Proceedings of the Third International Symposium on Non-Metallic (FRP) Reinforcement for Concrete Structures (FRPRCS-3)*, 1997, pp. 227–334.
184. F. Larsson, "The effect of ultraviolet light on mechanical properties of Kevlar 49 composites," *J. Reinf. Plast. Compos.*, vol. 5, no. 1, pp. 19–22, 1986, doi: 10.1177/073168448600500103.
185. CSA, *Design and Construction of Building Structures with Fibre-Reinforced Polymers (CSA S806-12 (R2017))*. Canadian Standards Association, Mississauga, ON, Canada, 2012.
186. J. Zhao, G. Cai, L. Cui, A. Si Larbi, and K. D. Tsavdaridis, "Deterioration of basic properties of the materials in FRP-strengthening RC structures under ultraviolet exposure," *Polymers (Basel).*, vol. 9, no. 402, 2017, doi: 10.3390/polym9090402.
187. B. D. Agarwal and L. J. Broutman, *Analysis and Performance of Fiber Composites*, 2nd ed. New York: Wiley, 1990.
188. L. C. Bank, *Properties of FRP Reinforcements for Concrete*. Elsevier B.V., 1993.
189. L. A. Bisby, B. K. Williams, M. F. Green, and V. K. Kodur, "Studies on the fire behaviour of FRP reinforced and/or strengthened concrete members," in *2nd International Conference on the Durability of Fibre Reinforced Polymer Composites for Construction*, 2002, pp. 405–417.
190. H. Hajiloo and M. F. Green, "Bond strength of GFRP reinforcing bars at high temperatures with implications for performance in fire," *J. Compos. Constr.*, vol. 22, no. 6, p. 04018055, 2018, doi: 10.1061/(asce)cc.1943-5614.0000897.
191. F. Sha69rifianjazi *et al.*, "Fibre-reinforced polymer reinforced concrete members under elevated temperatures: A review on structural performance," *Polymers (Basel).*, vol. 14, no. 3, p. 472, 2022, doi: 10.3390/polym14030472.
192. Y. C. Wang, P. M. H. Wong, and V. Kodur, "Mechanical properties of fibre reinforced polymer reinforcing bars at elevated temperatures," in *SFPE/ASCE Specialty Conference: Designing Structures for Fire*, 2003, pp. 183–192.
193. V. Kodur and D. Baingo, "Fire resistance of FRP reinforced concrete slabs," in *Interflam '99, 8th International Fire Science and Engineering Conference, Edinbugh, Scotland, UK*, 1999, pp. 928–937.
194. Y. C. Wang and V. Kodur, "Variation of strength and stiffness of fibre reinforced polymer reinforcing bars with temperature," *Cem. Concr. Compos.*, vol. 27, no. 9–10, pp. 864–874, 2005, doi: 10.1016/j.cemconcomp.2005.03.012.
195. A. Weber, "Fire-resistance tests on composite rebars," in *Proceedings of the 4th International Conference on FRP Composites in Civil Engineering, CICE 2008*, 2008, pp. 22–24.

196. V. Kodur and L. A. Bisby, "Evaluation of fire endurance of concrete slabs reinforced with fiber-reinforced polymer bars," *J. Struct. Eng.*, vol. 131, no. 1, pp. 34–43, 2005, doi: 10.1061/(asce)0733-9445(2005)131:1(34).
197. A. Abbasi and P. J. Hogg, "Fire testing of concrete beams with fibre reinforced plastic rebar," *Adv. Polym. Compos. Struct. Appl. Constr.*, pp. 445–456, 2004, doi: 10.1533/9781845690649.5.445.
198. E. Nigro, G. Cefarelli, A. Bilotta, G. Manfredi, and E. Cosenza, "Fire resistance of concrete slabs reinforced with FRP bars. Part I: Experimental investigations on the mechanical behavior," *Compos. Part B Eng.*, vol. 42, no. 6, pp. 1739–1750, 2011, doi: 10.1016/j.compositesb.2011.02.025.
199. E. Nigro, G. Cefarelli, A. Bilotta, G. Manfredi, and E. Cosenza, "Fire resistance of concrete slabs reinforced with FRP bars. Part II: Experimental results and numerical simulations on the thermal field," *Compos. Part B Eng.*, vol. 42, no. 6, pp. 1751–1763, 2011, doi: 10.1016/j.compositesb.2011.02.026.
200. E. Nigro, A. Bilotta, G. Cefarelli, G. Manfredi, and E. Cosenza, "Performance under fire situations of concrete members reinforced with FRP rods: Bond models and design nomograms," *J. Compos. Constr.*, vol. 16, no. 4, pp. 395–406, 2012, doi: 10.1061/(asce)cc.1943-5614.0000279.
201. M. M. Rafi, A. Nadjai, F. Ali, and H. C. Group, "Evaluation of thermal resistance of FRP reinforced concrete beams in fire," *J. Struct. Fire Eng.*, vol. 2, no. 2, pp. 91–107, 2011.
202. N. H. Albu-hassan and H. Al-thairy, "Experimental and numerical investigation on the behavior of hybrid concrete beams reinforced with GFRP bars after exposure to elevated temperature," *Structures*, vol. 28, no. September, pp. 537–551, 2020, doi: 10.1016/j.istruc.2020.08.079.
203. K. Protchenko and E. Szmigiera, "Post-fire characteristics of concrete beams reinforced with hybrid FRP bars," *Materials*, vol. 13, no. 5, pp. 1–15, 2020.
204. J. Prokeš, I. Rozsypalová, F. Girgle, P. Daněk, and P. Štěpánek, "Effects of elevated temperature on the behaviour of concrete beams reinforced with fiber reinforced polymers," 2021, doi: 10.1088/1757-899X/1039/1/012008.
205. M. Deuring, *Brandversuche an Nachtraglich Verstarkten Tragern aus Beton. Research Report EMPA No. 148,795*. Swiss Federal Laboratories for Materials Testing and Research, Dubendorf, Switzerland, 1994.
206. H. Blontrock, L. Taerwe, and P. Vandevelde, "Fire tests on concrete beams strengthened with fibre composite laminates," 2000.
207. L. A. Bisby, M. F. Green, and V. K. R. Kodur, "Response to fire of concrete structures that incorporate FRP," *Prog. Struct. Eng. Mater.*, vol. 7, no. 3, pp. 136–149, 2005, doi: 10.1002/pse.198.
208. B. Williams, V. Kodur, M. Green, and L. Bisby, "Fire endurance of fiber-reinforced polymer strengthened concrete T-Beams," *ACI Struct. J.*, vol. 105, pp. 60–67, Jan. 2008.
209. M. A. Aiello, F. Fofacci, and A. Nanni, "Effects of thermal loads on concrete cover of FRP reinforced elements: Theoretical and experimental analysis," *ACI Mater. J.*, vol. 98, no. 4, pp. 332–339, 2001.
210. S. M. Homam and S. A. Sheikh, "Durability of fibre reinforced polymers used in concrete structures S.," in *Proc. 3rd International Conference on Advanced Materials in Bridges and Structures*, 2000, pp. 751–758.
211. R. Koller, S. Chang, and Y. Xi, "Fiber-reinforced polymer bars under freeze-thaw cycles and different loading rates," *J. Compos. Mater.*, vol. 41, no. 1, pp. 5–25, 2007, doi: 10.1177/0021998306063154.

212. H. Bellakehal, A. Zaidi, R. Masmoudi, and M. Bouhicha, "Combined effect of sustained load and freeze-thaw cycles on one-way concrete slabs reinforced with glass fibre – reinforced polymer," *Can. J. Civ. Eng.*, vol. 40, no. 11, pp. 1060–1067, 2013, doi: 10.1139/cjce-2012-0514.
213. F. Yan, Z. Lin, and M. Yang, "Bond mechanism and bond strength of GFRP bars to concrete: A review," *Compos. Part B Eng.*, vol. 98, pp. 56–69, 2016, doi: 10.1016/j.compositesb.2016.04.068.
214. M. Mashima and K. Iwamoto, "Bond characteristics of FRP rod and concrete after freezing and Thawing deterioration," *ACI Symp. Publ.*, vol. 138, no. 4, pp. 51–70, 1993, doi: 10.14359/10034.
215. J. P. Won and C. G. Park, "Effect of environmental exposure on the mechanical and bonding properties of hybrid FRP reinforcing bars for concrete structures," *J. Compos. Mater.*, vol. 40, no. 12, pp. 1063–1076, 2005, doi: 10.1177/0021998305057362.
216. K. Laoubi, E. El-Salakawy, and B. Benmokrane, "Creep and durability of sand-coated glass FRP bars in concrete elements under freeze/thaw cycling and sustained loads," *Cem. Concr. Compos.*, vol. 28, no. 10, pp. 869–878, 2006, doi: 10.1016/j.cemconcomp.2006.07.014.
217. J. Alves, A. El-Ragaby, and E. El-Salakawy, "Durability of GFRP bars' bond to concrete under different loading and environmental conditions," *J. Compos. Constr.*, vol. 15, no. 3, pp. 249–262, 2011, doi: 10.1061/(asce)cc.1943-5614.0000161.
218. F. Nardone, M. Di Ludovico, F. J. De Caso Y Basalo, A. Prota, and A. Nanni, "Tensile behavior of epoxy based FRP composites under extreme service conditions," *Compos. Part B Eng.*, vol. 43, no. 3, pp. 1468–1474, 2012, doi: 10.1016/j.compositesb.2011.08.042.
219. V. M. Karbhari and L. Zhao, "Issues related to composite plating and environmental exposure effects on composite-concrete interface in external strengthening," *Compos. Struct.*, vol. 40, no. 3–4, pp. 293–304, 1997, doi: 10.1016/S0263-8223(98)00031-2.
220. P. Mukhopadhyaya, R. N. Swamy, and C. J. Lynsdale, "Influence of aggressive exposure conditions on the behaviour of adhesive bonded concrete-GFRP joints," *Constr. Build. Mater.*, vol. 12, no. 8, pp. 427–446, 1998, doi: 10.1016/S0950-0618(98)00030-0.
221. M. Green, K. Soudki, and M. Johnson, "Freeze-Thaw behaviour of reinforced concrete beams strengthened by fibre reinforced plastic sheets," 1997.
222. S. M. Homam and S. A. Sheikh, "Durability of fiber reinforced polymers used in concrete structures," in *Proceedings of the 3rd International Conference on Advanced Materials in Bridges and Structures, Ottawa, Canada*, 2000, pp. 751–758.
223. M. F. Green, A. J. S. Dent, and L. A. Bisby, "Effect of freeze-thaw cycling on the behaviour of reinforced concrete beams strengthened in flexure with fibre reinforced polymer sheets," *Can. J. Civ. Eng.*, vol. 30, no. 6, pp. 1081–1088, 2003, doi: 10.1139/l03-059.
224. K. V. Subramaniam, M. Ali-Ahmad, and M. Ghosn, "Freeze-thaw degradation of FRP-concrete interface: Impact on cohesive fracture response," *Eng. Fract. Mech.*, vol. 75, no. 13, pp. 3924–3940, 2008, doi: 10.1016/j.engfracmech.2007.12.016.
225. S. M. Soliman, E. El-Salakawy, and B. Benmokrane, "Bond performance of near-surface-mounted FRP bars," *J. Compos. Constr.*, vol. 15, no. 1, pp. 103–111, 2011, doi: 10.1061/(asce)cc.1943-5614.0000150.
226. P. T. Curtis, "The fatigue behaviour of fibrous composite materials," *J. Strain Anal. Eng. Des.*, vol. 24, no. 4, pp. 235–244, 1989, doi: 10.1243/03093247V244235.
227. R. Adimi, H. Rahman, B. Benmokrane, and K. Kobayashi, "Effect of temperature and loading frequency on the fatigue life of a CFRP bar in concrete," in *Proceedings of the Second International Conference on Composites in Infrastructure (ICCI-98)*, 1998, vol. 2, pp. 203–210.

228. M. L. Porter, B. W. Hughes, and B. A. Barnes, *Non-Corrosive Tie Reinforcing and Dowel Bars for Highway Pavement Slabs*. Report No. HR-343. Iowa Highway Research Board and Iowa Department of Transportation, Ames, IA, 1993.
229. S. V. Kumar and H. V. S. GangaRao, "Fatigue response of concrete decks reinforced with FRP rebars," *J. Struct. Eng.*, vol. 124, no. 1, pp. 11–16, 1998, doi: 10.1061/(asce)0733-9445(1998)124:1(11).
230. A. El-Ragaby, E. El-Salakawy, and B. Benmokrane, "Fatigue analysis of concrete bridge deck slabs reinforced with E-glass/vinyl ester FRP reinforcing bars," *Compos. Part B Eng.*, vol. 38, no. 5–6, pp. 703–711, 2007, doi: 10.1016/j.compositesb.2006.07.012.
231. X. Zhao, X. Wang, Z. Wu, and Z. Zhu, "Fatigue behavior and failure mechanism of basalt FRP composites under long-term cyclic loads," *Int. J. Fatigue*, vol. 88, pp. 58–67, 2016, doi: 10.1016/j.ijfatigue.2016.03.004.
232. Y. Shao, K. Okubo, T. Fujii, O. Shibata, and Y. Fujita, "Effect of matrix properties on the fatigue damage initiation and its growth in plain woven carbon fabric vinylester composites," *Compos. Sci. Technol.*, vol. 104, pp. 125–135, 2014, doi: 10.1016/j.compscitech.2014.09.010.
233. E. Ferrier, D. Bigaud, P. Hamelin, L. Bizindavyi, and K. W. Neale, "Fatigue reliability of external bonded CFRP used for concrete structure," *Mater. Struct.*, vol. 38, no. 275, pp. 39–46, 2005.
234. E. Ferrier, D. Bigaud, J. C. Clément, and P. Hamelin, "Fatigue-loading effect on RC beams strengthened with externally bonded FRP," *Constr. Build. Mater.*, vol. 25, no. 2, pp. 539–546, 2011, doi: 10.1016/j.conbuildmat.2010.07.035.
235. X. H. Zheng, P. Y. Huang, G. M. Chen, and X. M. Tan, "Fatigue behavior of FRP-concrete bond under hygrothermal environment," *Constr. Build. Mater.*, vol. 95, pp. 898–909, 2015, doi: 10.1016/j.conbuildmat.2015.07.172.
236. Y. H. Mugahed Amran, R. Alyousef, R. S. M. Rashid, H. Alabduljabbar, and C. C. Hung, "Properties and applications of FRP in strengthening RC structures: A review," *Structures*, vol. 16, no. July, pp. 208–238, 2018, doi: 10.1016/j.istruc.2018.09.008.
237. F. M. Monticeli, A. K. dos Reis, R. M. Neves, L. F. de P. Santos, E. C. Botelho, and H. L. Ornaghi, "Statistical analysis of creep behavior in thermoset and thermoplastic composites reinforced with carbon and glass fibers," *J. Strain Anal. Eng. Des.*, vol. 56, no. 7, pp. 452–461, 2021, doi: 10.1177/0309324720976637.
238. N. Ando, H. Matsukawa, A. Hattori, and A. Mashima, "Experimental studies on the long-term tensile properties of FRP tendons," *Proc. Third Int. Symp. Non-Metallic Reinf. Concr. Struct.*, vol. 2, pp. 203–210, 1997.
239. Y. A. Al-Salloum and T. H. Almusallam, "Creep effect on the behavior of concrete beams reinforced with GFRP bars subjected to different environments," *Constr. Build. Mater.*, vol. 21, no. 7, pp. 1510–1519, 2007, doi: 10.1016/j.conbuildmat.2006.05.008.
240. T. Yamaguchi, Y. Kato, T. Nishimura, and T. Uomoto, "Creep rupture of FRP rods made of aramid, carbon and glass fibers," in *Proceedings of the Third International Symposium on Non-metallic (FRP) Reinforcement for Concrete Structures (FRPRCS-3)*, Jan. 1997, vol. 2, pp. 179–186.
241. H. Seki, K. Sekijima, and T. Konno, "Test method on creep of continuous fiber reinforcing materials," in *Proceedings of the Third International Symposium on Non-Metallic (FRP) Reinforcement for Concrete Structures (FRPRCS-3)*, 1997, pp. 195–202.
242. H. Saadatmanesh and M. R. Ehsani, "Relaxation, creep, and fatigue behavior of carbon fiber-reinforced plastic tendons," *ACI Mater. J.*, vol. 96, no. 2, pp. 143–153, 1999.
243. B. Benmokrane, V. L. Brown, K. Mohamed, A. Nanni, M. Rossini, and C. Shield, "Creep-rupture limit for GFRP bars subjected to sustained loads," *J. Compos. Constr.*, vol. 23, no. 6, pp. 1–7, 2019, doi: 10.1061/(ASCE)CC.1943-5614.0000971.

244. I. Vilanova, M. Baena, L. Torres, and C. Barris, "Experimental study of bond-slip of GFRP bars in concrete under sustained loads," *Compos. Part B Eng.*, vol. 74, pp. 42–52, 2015, doi: 10.1016/j.compositesb.2015.01.006.
245. K. K. Choi, P. Meshgin, and M. M. Reda Taha, "Shear creep of epoxy at the concrete-FRP interfaces," *Compos. Part B Eng.*, vol. 38, no. 5–6, pp. 772–780, 2007, doi: 10.1016/j.compositesb.2006.10.003.
246. E. Ferrier, L. Michel, B. Jurkiewiez, and P. Hamelin, "Creep behavior of adhesives used for external FRP strengthening of RC structures," *Constr. Build. Mater.*, vol. 25, no. 2, pp. 461–467, 2011, doi: 10.1016/j.conbuildmat.2010.01.002.
247. G. Al Chami, M. Thériault, and K. W. Neale, "Creep behaviour of CFRP-strengthened reinforced concrete beams," *Constr. Build. Mater.*, vol. 23, no. 4, pp. 1640–1652, 2009, doi: 10.1016/j.conbuildmat.2007.09.006.
248. J. Bai, *Advanced Fibre-Reinforced Polymer (FRP) Composites for Structural Applications*. Woodhead Publishing, 2013.
249. S. A. Grammatikos, B. Zafari, M. C. Evernden, J. T. Mottram, and J. M. Mitchels, "Moisture uptake characteristics of a pultruded fibre reinforced polymer flat sheet subjected to hot/wet aging," *Polym. Degrad. Stab.*, vol. 121, pp. 407–419, 2015, doi: 10.1016/j.polymdegradstab.2015.10.001.
250. ASTM D5117, *Standard Test Method for Dye Penetration of Solid Fiberglass Reinforced Pultruded*. ASTM International, West Conshohocken, PA, 2017.
251. J. D. Lee, *Plastic Shrinkage Cracking Testing Report*. Project Rep. No. 99-09552, Twin City Testing Corp., Twin City, MN, 1999.
252. Y. Shao and A. Mirmiran, "Control of plastic shrinkage cracking of concrete with carbon fiber-reinforced polymer grids," *J. Mater. Civ. Eng.*, vol. 19, no. 5, 2007, doi: 10.1061/(ASCE)0899-1561(2007)19.
253. ICC-ES, *AC-32. Acceptance Criteria for Concrete with Synthetic Fibers*. International Council of Codes Evaluation Services. Los Angeles, CA, 2003.
254. A. Nanni, *Fiber-Reinforced-Plastic (FRP) Reinforcement for Concrete Structures: Properties and Applications*. Developments in Civil Engineering, 1993.
255. H. V. S. GangaRao and P. V. Vijay, "Design of concrete members reinforced with GFRP bars," in *Proceedings of the Third International Symposium on Non-Metallic (FRP) Reinforcement for Concrete Structures (FRPRCS-3)*, 1997, pp. 143–150.
256. M. Theriaule and B. Benmokrane, "Effects of Frp reinforcement ratio and concrete strength," *J. Compos. Constr.*, vol. 2, no. February, pp. 7–16, 1998.
257. C. Barris, L. Torres, M. Baena, K. Pilakoutas, and M. Guadagnini, "Serviceability limit state of FRP RC beams," *Adv. Struct. Eng.*, vol. 15, no. 4, pp. 653–664, 2012, doi: 10.1260/1369-4332.15.4.653.
258. AASHTO, *Guide Specifications for the Design of Concrete Bridge Beams Prestressed with Carbon Fiber-Reinforced Polymer (CFRP) Systems*, 1st ed. American Association of State Highway and Transportation Officials, Washington, DC, 2018.
259. ACI, *Building Code Requirements for Structural Concrete and Commentary (ACI 318-19)*. American Concrete Institute, Farmington Hills, MI, 2019.
260. AASHTO, *LRFD Bridge Design Specifications*, 9th ed. American Association of State Highway and Transportation Officials, Washington, DC, 2020.
261. B. Behnam and C. Eamon, "Reliability-based design optimization of concrete flexural members reinforced with ductile FRP bars," *Constr. Build. Mater.*, vol. 47, pp. 942–950, 2013, doi: 10.1016/j.conbuildmat.2013.05.101.
262. L. A. Bisby and V. Kodur, "Evaluating the fire endurance of concrete slabs reinforced with FRP bars: Considerations for a holistic approach," *Compos. Part B Eng.*, vol. 38, no. 5–6, pp. 547–558, 2007, doi: 10.1016/j.compositesb.2006.07.013.

263. CSA S6:19, *Canadian Highway Bridge Design Code (CSA S6:19)*. Canadian Standards Association, Mississauga, ON, Canada, 2019.
264. K. Takahashi *et al.*, "In-situ observation of interfacial debonding between a single carbon fiber and epoxy matrix by synchrotron radiation X-ray nano-imaging," 2022, [Online]. Available: https://doi.org/10.21203/rs.3.rs-1314626/v1.
265. V. M. Karbhari, H. Kaiser, R. Navada, K. Ghosh, and L. Lee, "Methods for detecting defects in composite rehabilitated concrete structures," 2005.
266. P. Malla, S. S. Khedmatgozar Dolati, J. D. Ortiz, A. Mehrabi, and A. Nanni, "Damage and defects in fiber-reinforced polymer reinforced and strengthened concrete elements," *J. Compos. Constr.*, vol. 27, no. 4, p. 04023035, 2023.
267. J. S. Jung, J. W. Jeong, and K. S. Lee, "Structural performance degradation of corrosion-damaged reinforced concrete beams based on finite element analysis," *Appl. Sci.*, vol. 12, no. 4, 2022, doi: 10.3390/app12042090.
268. B. Shafei, "Assessment of bridge decks with Glass Fiber Reinforced Polymer (GFRP) reinforcement," no. March, 2023.
269. C. Barris, L. Torres, A. Turon, M. Baena, and A. Catalan, "An experimental study of the flexural behaviour of GFRP RC beams and comparison with prediction models," *Compos. Struct.*, vol. 91, no. 3, pp. 286–295, 2009, doi: 10.1016/j.compstruct.2009.05.005.
270. A. B. Mehrabi and S. Farhangdoust, "NDT methods applicable to health monitoring of ABC closure joints," 2019. [Online]. Available: https://abc-utc.fiu.edu/wp-content/uploads/sites/52/2019/04/2013-C3-FIU04-NDT-FinalReport.pdf.
271. Y. Chang, Y. Wang, M. Wang, Z. Zhou, and J. Ou, "Bond durability and degradation mechanism of GFRP bars in seawater sea-sand concrete under the coupling effect of seawater immersion and sustained load," *Constr. Build. Mater.*, vol. 307, no. August, p. 124878, 2021, doi: 10.1016/j.conbuildmat.2021.124878.
272. F. Riccitelli, A. Mehrabi, M. Abedin, S. Farhangdoust, and S. S. Khedmatgozar Dolati, "Performance of existing Abc projects: Inspection case studies," *Accel. Bridg. Constr. Univ. Transp. Cent. Miami, FL, USA*, 2020.
273. C. J. Hellier, *Handbook of Nondestructive Evaluation*, Third Edit. McGraw-Hil Education, 2020.
274. S. S. Khedmatgozar Dolati, N. Caluk, A. Mehrabi, and S. S. Khedmatgozar Dolati, "Non-destructive testing applications for steel bridges," *Appl. Sci.*, vol. 11, no. 20, p. 9757, 2021.
275. S. S. K. Dolati, A. Mehrabi, S. S. K. Dolati, and N. Caluk, "NDT methods for damage detection in steel bridges," in *Health Monitoring of Structural and Biological Systems XVI*, 2022, vol. 12048, pp. 385–394.
276. W. Li, S. C. M. Ho, D. Patil, and G. Song, "Acoustic emission monitoring and finite element analysis of debonding in fiber-reinforced polymer rebar reinforced concrete," *Struct. Heal. Monit.*, vol. 16, no. 6, pp. 674–681, 2017, doi: 10.1177/1475921716678922.
277. A. Mirmiran and Y. Wei, "Damage assessment of FRP-encased concrete using ultrasonic pulse velocity," pp. 126–135, 2001.
278. S. Ettouney and M. Alampalli, *Infrastructure Health in Civil Engineering*. Boca Raton, FL: CRC Press, 2011.
279. N. Yazdani, E. C. Garcia, and M. Riad, *Field Assessment of Concrete Structures Rehabilitated with FRP*. Elsevier Ltd, 2018.
280. P. Cawley and R. D. Adams, "The mechanics of the coin-tap method of non-destructive testing," *J. Sound Vib.*, vol. 122, no. 2, pp. 299–316, 1988, doi: 10.1016/S0022-460X(88)80356-0.
281. R. I. Mackie and A. E. Vardy, "Applying the coin-tap test to adhesives in civil engineering: A numerical study," *Int. J. Adhes. Adhes.*, vol. 10, no. 3, pp. 215–220, 1990, doi: 10.1016/0143-7496(90)90106-8.

282. H. Kaiser and V. M. Karbhari, "Non-destructive testing techniques for FRP rehabilitated concrete - I: A critical review," *Int. J. Mater. Prod. Technol.*, vol. 21, no. 5, pp. 349–384, 2004, doi: 10.1504/IJMPT.2004.004996.
283. A. S. Wheeler, "Nondestructive evaluation of concrete bridge columns rehabilitiated with fiber reinforced polymers using digital tap hammer and infrared thermography nondestructive evaluation of concrete bridge columns," 2018.
284. U. B. Halabe, R. M. Joshi, and H. V. S. Gangarao, "Nondestructive testing of FRP composite structural components and FRP rehabilitated bridge using digital tap testing," vol. 7, no. 1, pp. 11477–11482, 2020.
285. S. Khanal, "Review of modern nondestructive testing techniques for civil infrastructure," West Virginia University, 2020.
286. M. Riad, "Quantitative Non-Destructive Evaluation (NDE) of FRP laminate-concrete bond strength," The University of Texas at Arlington, 2017.
287. M. Ekenel and J. J. Myers, "Nondestructive evaluation of RC structures strengthened with FRP laminates containing near-surface defects in the form of delaminations," *Sci. Eng. Compos. Mater.*, vol. 14, no. 4, pp. 299–315, 2007, doi: 10.1515/secm.2007.14.4.299.
288. F. Taillade, M. Quiertant, K. Benzarti, J. Dumoulin, and C. Aubagnac, "Nondestructive evaluation of FRP strengthening systems bonded on RC structures using pulsed stimulated infrared thermography," in *Infrared Thermography, Chapter 9*, 2012, pp. 193–208.
289. ACI, "ACI 228.2R-13: Report on nondestructive test methods for evaluation of concrete in structures," 2013. doi: 10.14359/51686889.
290. N. J. Carino, M. Sansalone, and N. N. Hsu, "Flaw detection in concrete by frequency spectrum analysis of impact-echo waveforms," *Int. Adv. Nondestruct. Test.*, vol. 12, pp. 117–146, 1986.
291. Y. Dong and F. Ansari, "Non-destructive testing and evaluation (NDT/NDE) of civil structures rehabilitated using fiber reinforced polymer (FRP) composites," *Serv. Life Estim. Ext. Civ. Eng. Struct.*, pp. 193–222, 2011, doi: 10.1533/9780857090928.2.193.
292. O. Gunes Dr, "Failure modes in structural applications of fiber-reinforced polymer (FRP) composites and their prevention," in *Developments in Fiber-Reinforced Polymer (FRP) Composites for Civil Engineering*, edited by N. Uddin, pp. 115–147. Cambridge, UK: Woodhead Publishing Limited, 2013.
293. N. H. Maerz and G. Galecki, "Preservation of Missouri transportation infrastructures: Validation of FRP composite technology," 2008.
294. N. H. Maerz, G. Galecki, and A. Nanni, "Experimental non-destructive testing of FRP materials, installation, and performance, Dallas County Bridge, Missouri, USA," 2004.
295. F. Schubert and B. Köhler, "Ten lectures on impact-echo," *J. Nondestruct. Eval.*, vol. 27, no. 1–3, pp. 5–21, 2008, doi: 10.1007/s10921-008-0036-2.
296. V. M. Malhotra and N. J. Carino, *Handbook on Nondestructive Testing of Concrete,* 2nd ed. Boca Raton, FL: CRC Press, 2004.
297. S. Farhangdoust and A. Mehrabi, "ABC-UTC guide for: Selection of NDT methods applicable to health monitoring," 2020.
298. A. Mehrabi and S. Farhangdoust, "NDT methods applicable to health monitoring of ABC closure joints," Accelerated Bridge Construction University Transportation Center (ABC-UTC), 2019.
299. D. M. Pozar, *Microwave Engineering*, 4th ed. John Wiley & Sons, 2012.
300. M. Q. Feng, G. Roqueta, and L. Jofre, "Non-destructive evaluation (NDE) of composites: Microwave techniques," in *Non-Destructive Evaluation (NDE) of Polymer Matrix Composites*, edited by V. M. Karbhari, pp. 574–616. Cambridge, UK: Woodhead Publishing Limited, 2013.

301. S. I. Ganchev, N. Qaddoumi, E. Ranu, and R. Zoughi, "Microwave detection optimization of disbond in layered dielectrics with varying thickness," *IEEE Trans. Instrum. Meas.*, vol. 44, no. 2, pp. 326–328, 1995, doi: 10.1109/19.377843.
302. B. Akuthota, D. Hughes, R. Zoughi, J. Myers, and A. Nanni, "Near-field microwave detection of disbond in carbon fiber reinforced polymer composites used for strengthening cement-based structures and disbond repair verification," *J. Mater. Civ. Eng.*, vol. 16, no. 6, pp. 540–546, 2004, doi: 10.1061/(asce)0899-1561(2004)16:6(540).
303. M. Q. Feng, F. De Flaviis, and Y. J. Kim, "Use of microwaves for damage detection of fiber reinforced polymer-wrapped concrete structures," *J. Eng. Mech.*, vol. 128, no. 2, pp. 172–183, 2002, doi: 10.1061/(asce)0733-9399(2002)128:2(172).
304. S. Kharkovsky, A. C. Ryley, V. Stephen, and R. Zoughi, "Dual-polarized microwave near-field reflectometer for non-invasive inspection of carbon fiber reinforced polymer (CFRP) strengthened structures," *Conf. Rec. – IEEE Instrum. Meas. Technol. Conf.*, no. April, pp. 2108–2111, 2006, doi: 10.1109/IMTC.2006.236695.
305. S. Kharkovsky and R. Zoughi, "Application of near-field microwave and millimeter wave nondestructive testing for evaluation of fiber breakage and orientation evaluation in CFRP composite patches," *AIP Conf. Proc.*, vol. 760, pp. 554–561, 2005, doi: 10.1063/1.1916724.
306. J. Li and C. Liu, "Noncontact detection of air voids under glass epoxy jackets using a microwave system," *Subsurf. Sens. Technol. Appl.*, vol. 2, no. 4, pp. 411–423, 2001, doi: 10.1023/A:1013221118462.
307. V. Stephen, S. Kharkovsky, J. Nadakuduti, and R. Zoughi, "Microwave field measurement of delaminations in CFRP concrete members in a bridge," 2004.
308. S. S. Dutta, "Nondestructive evaluation of FRP wrapped concrete cylinders using infrared thermography and groud penetrating radar," West Virginia University, 2006.
309. T. Y. Yu and O. Büyüköztürk, "A far-field airborne radar NDT technique for detecting debonding in GFRP-retrofitted concrete structures," *NDT E Int.*, vol. 41, no. 1, pp. 10–24, 2008, doi: 10.1016/j.ndteint.2007.07.002.
310. A. Benedetto and L. Pajewski, eds., *Civil Engineering Applications of Ground Penetrating Radar*. Cham, Switzerland: Springer, 2015.
311. P. Kot, M. Muradov, M. Gkantou, G. S. Kamaris, K. Hashim, and D. Yeboah, "Recent advancements in non-destructive testing techniques for structural health monitoring," *Appl. Sci.*, vol. 11, no. 6, 2021, doi: 10.3390/app11062750.
312. Ł. Drobiec, R. Jasiński, and W. Mazur, "The use of non-destructive methods to detect non-metallic reinforcement in concrete and masonry," 2019, doi: 10.20944/preprints201901.0021.v1.
313. D. Jackson, M. Islam, and S. Alampalli, "Feasibility of evaluating the performance of fiber reinforced plastic (FRP) wrapped reinforced concrete columns using ground penetrating RADAR (GPR) and infrared (IR) thermography techniques," in *Structural Materials Technology IV-An NDT Conference*, 2000, pp. 390–395.
314. C. L. C. Hing and U. B. Halabe, "Nondestructive testing of GFRP bridge decks using ground penetrating radar and infrared thermography," *J. Bridg. Eng.*, vol. 15, no. 4, pp. 391–398, 2010, doi: 10.1061/(asce)be.1943-5592.0000066.
315. N. Yazdani, E. Beneberu, and M. Riad, "Nondestructive evaluation of FRP-concrete interface bond due to surface defects," *Adv. Civ. Eng.*, vol. 2019, 2019, doi: 10.1155/2019/2563079.
316. T. Kundu, M. Ehsani, K. I. Maslov, and D. Guo, "C-scan and L-scan generated images of the concrete/GFRP composite interface," *NDT & E International*, vol. 32, pp. 61–69, 1999.

317. F. Bastianini, A. Di Tommaso, and G. Pascale, "Ultrasonic non-destructive assessment of bonding defects in composite structural strengthenings," *Compos. Struct.*, vol. 53, pp. 463–467, 2001.
318. E. La Malfa Ribolla, M. R. Hajidehi, P. Rizzo, G. F. Scimemi, A. Spada, and G. Giambanco, "Ultrasonic inspection for the detection of debonding in CFRP-reinforced concrete," *Struct. Infrastruct. Eng.*, vol. 14, no. 6, pp. 807–816, 2017, doi: 10.1080/15732479.2017.1384843.
319. G. Concu and N. Trulli, "Direct and semi-direct ultrasonic testing for quality control of FRC-concrete adhesion," *Structures*, vol. 32, no. November 2020, pp. 54–64, 2021, doi: 10.1016/j.istruc.2021.02.061.
320. S. Gholizadeh, "A review of non-destructive testing methods of composite materials," *Procedia Struct. Integr.*, vol. 1, pp. 50–57, 2016, doi: 10.1016/j.prostr.2016.02.008.
321. P. J. Shull, *Nondestructive Evaluation Theory, Techniques, and Applications*. Marcel Dekker, Inc., 2001.
322. H. Taheri and A. A. Hassen, "Nondestructive ultrasonic inspection of composite materials: A comparative advantage of phased array ultrasonic," 2019.
323. H. Taheri, F. Delfanian, and J. Du, "Acoustic emission and ultrasound phased array technique for composite material evaluation," pp. 1–8, 2016.
324. A. S. Boychuk, A. S. Generalov, and A. V. Stepanov, "Nondestructive testing of FRP by using phased array ultrasonic technology," *12th Int. Conf. Slov. Soc. Non-Destructive Test. Appl. Contemp. Non-Destructive Test. Eng. ICNDT 2013 – Conf. Proc.*, pp. 51–55, 2013.
325. C. Meola, S. Boccardi, G. M. Carlomagno, N. D. Boffa, E. Monaco, and F. Ricci, "Nondestructive evaluation of carbon fibre reinforced composites with infrared thermography and ultrasonics," *Compos. Struct.*, vol. 134, pp. 845–853, 2015, doi: 10.1016/j.compstruct.2015.08.119.
326. A. S. Boychuk, A. S. Generalov, and A. V. Stepanov, "CFRP structural health monitoring by ultrasonic phased array technique," *7th Eur. Work. Struct. Heal. Monit. EWSHM 2014 – 2nd Eur. Conf. Progn. Heal. Manag. Soc.*, pp. 2206–2211, 2014.
327. T. W. Ryan, J. EricMann, Z. M. Chill, and B. T. Ott, "Bridge inspector's reference manual," *Fhwa*, vol. BIRM 1, p. 1020, 2012, [Online]. Available: www.nhi.fhwa.dot.gov.
328. U. Galietti, V. Luprano, S. Nenna, L. Spagnolo, and A. Tundo, "Non-destructive defect characterization of concrete structures reinforced by means of FRP," *Infrared Phys. Technol.*, vol. 49, no. 3 SPEC. ISS., pp. 218–223, 2007, doi: 10.1016/j.infrared.2006.06.008.
329. M. A. Starnes, N. J. Carino, and E. A. Kausel, "Preliminary thermography studies for quality control of concrete structures strengthened with fiber-reinforced polymer composites," *J. Mater. Civ. Eng.*, vol. 15, no. 3, pp. 266–273, 2003, doi: 10.1061/(asce)0899-1561(2003)15:3(266).
330. J. R. Brown and H. R. Hamilton, "Heating methods and detection limits for infrared thermography inspection of fiber-reinforced polymer composites," *ACI Mater. J.*, vol. 104, no. 5, pp. 481–490, 2007, doi: 10.14359/18904.
331. A. Mirmiran, M. Shahawy, and H. El Echary, "Acoustic emission monitoring of hybrid FRP-concrete columns," pp. 899–905, 1999.
332. R. S. Gostautas, G. Ramirez, R. J. Peterman, and D. Meggers, "Acoustic emission monitoring and analysis of glass fiber-reinforced composites bridge decks," *J. Bridg. Eng.*, vol. 10, no. 6, pp. 713–721, 2005, doi: 10.1061/(asce)1084-0702(2005)10:6(713).
333. A. Carpinteri, G. Lacidogna, and M. Paggi, "Acoustic emission monitoring and numerical modeling of FRP delamination in RC beams with non-rectangular cross-section," *Mater. Struct. Constr.*, vol. 40, no. 6, pp. 553–566, 2007, doi: 10.1617/s11527-006-9162-4.

334. S. Degala, P. Rizzo, K. Ramanathan, and K. A. Harries, "Acoustic emission monitoring of CFRP reinforced concrete slabs," *Constr. Build. Mater.*, vol. 23, no. 5, pp. 2016–2026, 2009.
335. A. R. Ganesan, "Holographic and laser speckle methods in non-destructive testing," *Proc. Natl. Semin. Exhib. Non-Destructive Eval.*, pp. 126–130, 2009, [Online]. Available: https://pdfs.semanticscholar.org/f4eb/9b46d271b9a7caa47e5f10d90e738bbbf6a1.pdf.
336. J. G. Chen, O. Buyukozturk, and R. W. Haupt, "Operational and defect parameters concerning the acoustic-laser vibrometry method for FRP-reinforced concrete," *NDT E Int.*, vol. 71, pp. 43–53, 2015, doi: 10.1016/j.ndteint.2014.12.006.
337. O. Büyüköztürk, R. Haupt, C. Tuakta, and J. Chen, "Remote detection of debonding in FRP-strengthened concrete structures using acoustic-laser technique," in *Nondestructive Testing of Materials and Structures*, 2013, vol. 6, pp. 19–24.
338. F. Taillade, M. Quiertant, K. Benzarti, C. Aubagnac, and E. Moser, "Non-destructive evaluation (NDE) of composites: Using shearography to detect bond defects," *Non-Destructive Eval. Polym. Matrix Compos. Tech. Appl.*, pp. 542–556, 2013, doi: 10.1533/9780857093554.4.542.
339. F. Taillade, M. Quiertant, K. Benzarti, C. Aubagnac, and E. Moser, "Shearography applied to the non destructive evaluation of bonded interfaces between concrete and CFRP overlays: From the laboratory to the field," *Eur. J. Environ. Civ. Eng.*, vol. 15, no. 4, pp. 545–556, 2010, doi: 10.3166/EJECE.15.545-556.
340. L. Yang, "Recent developments in digital shearography for nondestructive testing," *Mater. Eval.*, vol. 64, no. 7, pp. 704–709, 2006.
341. Q. Qiu and D. Lau, "A novel approach for near-surface defect detection in FRP-bonded concrete systems using laser reflection and acoustic-laser techniques," *Constr. Build. Mater.*, vol. 141, pp. 553–564, 2017, doi: 10.1016/j.conbuildmat.2017.03.024.
342. Q. Qiu and D. Lau, "Experimental evaluation on the effectiveness of acoustic-laser technique towards the FRP-bonded concrete system," *Struct. Heal. Monit. Insp. Adv. Mater. Aerospace, Civ. Infrastruct. 2015*, vol. 9437, p. 943705, 2015, doi: 10.1117/12.2086479.
343. J. G. Chen, R. W. Haupt, and O. Büyüköztürk, "Remote characterization of defects in FRP strengthened concrete using the acoustic-laser vibrometry method," 2013.
344. Y. K. Zhu, G. Y. Tian, R. S. Lu, and H. Zhang, "A review of optical NDT technologies," *Sensors*, vol. 11, no. 8, pp. 7773–7798, 2011, doi: 10.3390/s110807773.
345. D. Lau and Q. Qiu, "A review of nondestructive testing approaches using mechanical and electromagnetic waves," *Nondestruct. Charact. Monit. Adv. Mater. Aerospace, Civ. Infrastruct. 2016*, vol. 9804, p. 98042E, 2016, doi: 10.1117/12.2218739.
346. B. Wan, *Using Fiber-Reinforced Polymer (FRP) Composites in Bridge Construction and Monitoring their Performance: An Overview*. Woodhead Publishing Limited, 2014.
347. B. M. A. J. Mariscotti, F. Jalinoos, T. Frigerio, M. Ruffolo, and P. Thieberger, "Gamma-ray imaging for void and corrosion assessment safety and utility of the technology make it appropriate for field application," no. november, 2009.
348. S. Alampalli, "Field performance of an FRP slab bridge," *Compos. Struct.*, vol. 72, no. 4, pp. 494–502, 2006, doi: 10.1016/j.compstruct.2005.01.017.
349. H. Guan, V. M. Karbhari, and C. S. Sikorsky, "Long-term structural health monitoring system for a FRP composite highway bridge structure," *J. Intell. Mater. Syst. Struct.*, vol. 18, no. 8, pp. 809–823, 2007, doi: 10.1177/1045389X06073471.
350. O. Hag-elsafi, J. Kunin, S. Alampalli, and T. Conway, "Strengthening of route 378 bridge over Wynantskill creek in New York using FRP laminates," 2001.
351. O. Hag-elsafi, R. Lund, and S. Alampalli, "Strengthening of Church Street Bridge Pier Capbeam using bonded FRP composite plates: Strengthening and load testing," 2002.

352. R. K. Amineh, M. Ravan, and R. Sharma, "Nondestructive testing of nonmetallic pipes using wideband microwave measurements," *IEEE Trans. Microw. Theory Tech.*, vol. 68, no. 5, pp. 1763–1772, 2020.
353. Y. Shi, C. Zhang, R. Li, M. Cai, and G. Jia, "Theory and application of magnetic flux leakage pipeline detection," *Sensors (Switzerland)*, vol. 15, no. 12, pp. 31036–31055, 2015, doi: 10.3390/s151229845.
354. D. Li, J. Zhou, and J. Ou, "Damage, nondestructive evaluation and rehabilitation of FRP composite-RC structure: A review," *Constr. Build. Mater.*, vol. 271, p. 121551, 2021, doi: 10.1016/j.conbuildmat.2020.121551.
355. A. Azizinamini, "Non-destructive testing (NDT) of a segmental concrete bridge scheduled for demolition, with a focus on condition assessment and corrosion detection of internal tendons," 2017.
356. M. F. Ismail, M. T. H. Sultan, A. Hamdan, A. U. M. Shah, and M. Jawaid, "Low velocity impact behaviour and post-impact characteristics of kenaf/glass hybrid composites with various weight ratios," *J. Mater. Res. Technol.*, vol. 8, no. 3, pp. 2662–2673, 2019, doi: 10.1016/j.jmrt.2019.04.005.
357. A. Wronkowicz-Katunin, "A brief review on NDT&E methods for structural aircraft components," *Fatigue Aircr. Struct.*, vol. 2018, no. 10, pp. 73–81, 2018, doi: 10.2478/fas-2018-0007.
358. X. E. Gros and K. Takahashi, "Non-destructive characterization of delaminated areas at interfaces between plies in carbon fiber/epoxy laminates with Foucault currents," *Compos. Interfaces*, vol. 7, no. 3, pp. 177–192, 2000, doi: 10.1163/156855400300185298.
359. G. Mook, R. Lange, and O. Koeser, "Non-destructive characterisation of carbon-fibre-reinforced plastics by means of eddy-currents," *Compos. Sci. Technol.*, vol. 61, no. 6, pp. 865–873, 2001, doi: 10.1016/S0266-3538(00)00164-0.
360. R. Prakash, "Non-destructive testing of composites," *Composites*, vol. 11, no. 4, pp. 217–224, 1980, doi: 10.1016/0010-4361(80)90428-0.
361. M. Yumnam, H. Gupta, D. Ghosh, and J. Jaganathan, "Inspection of concrete structures externally reinforced with FRP composites using active infrared thermography: A review," *Constr. Build. Mater.*, vol. 310, no. May, p. 125265, 2021, doi: 10.1016/j.conbuildmat.2021.125265.
362. C. Z. Dong and F. N. Catbas, "A review of computer vision–based structural health monitoring at local and global levels," *Struct. Heal. Monit.*, vol. 20, no. 2, pp. 692–743, 2021, doi: 10.1177/1475921720935585.
363. Q. Qiu, "Imaging techniques for defect detection of fiber reinforced polymer-bonded civil infrastructures," *Struct. Control Heal. Monit.*, vol. 27, no. 8, pp. 1–22, 2020, doi: 10.1002/stc.2555.
364. N. Catbas, C. Z. Dong, O. Celik, and T. Khuc, "A vision for vision-based technologies for bridge health monitoring," *Maintenance, Safety, Risk, Manag. Life-Cycle Perform. Bridg. - Proc. 9th Int. Conf. Bridg. Maintenance, Saf. Manag. IABMAS 2018*, no. July, pp. 54–62, 2018, doi: 10.1201/9781315189390-4.
365. D. Ribeiro, R. Calçada, J. Ferreira, and T. Martins, "Non-contact measurement of the dynamic displacement of railway bridges using an advanced video-based system," *Eng. Struct.*, vol. 75, pp. 164–180, 2014.
366. H. Ahmed, H. M. La, and N. Gucunski, "Review of non-destructive civil infrastructure evaluation for bridges: State-of-the-art robotic platforms, sensors and algorithms habib," *Sensors*, vol. 20, no. 3954, pp. 1–38, 2020.
367. H. M. La, N. Gucunski, K. Dana, and S. H. Kee, "Development of an autonomous bridge deck inspection robotic system," *J. F. Robot.*, vol. 34, no. 8, pp. 1489–1504, 2017, doi: 10.1002/rob.21725.

368. S. T. Nguyen and H. M. La, "Development of a steel bridge climbing robot," *IEEE Int. Conf. Intell. Robot. Syst.*, pp. 1912–1917, 2019, doi: 10.1109/IROS40897.2019.8967748.
369. B. Lei, N. Wang, P. Xu, and G. Song, "New crack detection method for bridge inspection using UAV incorporating image processing," *J. Aerosp. Eng.*, vol. 31, no. 5, p. 04018058, 2018, doi: 10.1061/(asce)as.1943-5525.0000879.
370. J. Seo, L. Duque, and J. Wacker, "Drone-enabled bridge inspection methodology and application," *Autom. Constr.*, vol. 94, no. June, pp. 112–126, 2018, doi: 10.1016/j.autcon.2018.06.006.
371. K. D. S. Saman, M. Pranit, M. Armin, P. J. Ortiz, and N. Antonio, "Damage detection in concrete elements reinforced or strengthened with FRP and new methods for improving the detectability of FRP bars in concrete," in *Structures Congress 2025*, 2025, pp. 373–393.
372. P. Malla, S. S. K. Dolati, J. D. Ortiz, A. B. Mehrabi, A. N. Ding, and K. Dinh, "Applicability of available NDT methods for damage detection in concrete elements reinforced or strengthened with FRP," *Bridg. Struct.*, vol. 19, no. 4, pp. 149–164, 2023.
373. ASTM International, *Standard Test Method for Compressive Strength of Cylindrical Concrete Specimens.* ASTM C39/C39M-10. West Conshohocken, PA: ASTM International, 2010.
374. M. Ekenel, V. Stephen, J. J. Myers, and R. Zoughi, "Microwave NDE of reinforced concrete beams strengthened with CFRP laminates containing surface defects and tested under cyclic loading," 2004.
375. H. Kaiser, V. M. Karbhari, and C. Sikorsky, "Non-destructive testing techniques for FRP rehabilitated concrete ± II: An assessment," 2004.
376. C. T. Hsieh and Y. Lin, "Detecting debonding flaws at the epoxy-concrete interfaces in near-surface mounted CFRP strengthening beams using the impact-echo method," *NDT E Int.*, vol. 83, pp. 1–13, 2016, doi: 10.1016/j.ndteint.2016.05.001.
377. K. C. Crawford, "Non-destructive testing of FRP-structural systems applied to concrete bridges," in *Nondestructive Testing of Materials and Structures.* Dordrecht: Springer, 2013, pp. 835–840.
378. K. C. Crawford, "NDT evaluation of long-term bond durability of CFRP-structural systems applied to RC highway bridges," *Int. J. Adv. Struct. Eng.*, vol. 8, no. 2, pp. 161–168, Jun. 2016, doi: 10.1007/s40091-016-0120-2.
379. S. Kharkovsky, A. C. Ryley, V. Stephen, and R. Zoughi, "Dual-polarized near-field microwave reflectometer for noninvasive inspection of carbon fiber reinforced polymer-strengthened structures," *IEEE Trans. Instrum. Meas.*, vol. 57, no. 1, pp. 168–175, 2008, doi: 10.1109/TIM.2007.909497.
380. M. D. Navagato and R. M. Narayanan, "Microwave imaging of multilayered structures using ultrawideband noise signals," *NDT E Int.*, vol. 104, no. February, pp. 19–33, 2019, doi: 10.1016/j.ndteint.2019.02.009.
381. D. Hughes, M. Kazemi, K. Marler, R. Zoughi, J. Myers, and A. Nanni, "Microwave detection of delaminations between fiber reinforced polymer (FRP) composite and hardened cement paste," in *AIP Conference Proceedings*, 2002, pp. 512–519.
382. Y. J. Kim, F. De Flaviis, L. Jofre, and M. Q. Feng, "Microwave-based NDE of FRP-Jacketed concrete structures," in *International SAMPE Symposium and Exhibition*, 1999, pp. 439–450.
383. O. Büyüköztürk and T. Y. Yu, "Far-field radar NDT technique for detecting GFRP debonding from concrete," *Constr. Build. Mater.*, vol. 23, no. 4, pp. 1678–1689, Apr. 2009, doi: 10.1016/j.conbuildmat.2007.09.009.

384. M. Riad, N. Yazdani, Y. Almomani, and E. Gonzalez, "Utilization of Ground Penetrating Radar (GPR) in the non-destructive assessment of FRP laminate-concrete bond strength," in *Sustainable Civil Infrastructures*, 2019, pp. 244–267, doi: 10.1007/978-3-030-01932-7_20.
385. R. Franklin, U. B. Halabe, R. Lopez-Anido, and S. Kshirsagar, "Condition assessment of FRP wrapped concrete members using ultrasonics," *Nondestruct. Test. Eval.*, vol. 17, no. 1, pp. 59–78, 2001, doi: 10.1080/10589750108953102.
386. M. Ekenel, N. Galati, J. J. Myers, A. Nanni, and V. Godínez, "Acousto-ultrasonic technology for nondestructive evaluation of concrete bridge members strengthened by carbon fiber – Reinforced polymer," *Transp. Res. Rec.*, vol. 1928, no. 1, pp. 245–251, 2005.
387. C. V Subramanian, M. Thavasimuthu, P. Palanichamy, D. K. Bhattacharya, and B. Raj, "Evaluation of bond integrity in sandwiched structures by dry couplant ultrasonic technique," *NDT E Int.*, vol. 24, no. 1, pp. 29–31, 1991.
388. V. A. Luprano, A. Tundo, I. Angelo TATÌ, I. Ermanno Grinzato, S. Marinetti, and P. G. Bison, "Non destructive defect characterization in civil structures reinforced by means of FRP," 2006.
389. R. Helmerich *et al.*, "Near surface mounted CFRP-rods–Preparation of reinforcing measures and quality control by means of non-destructive testing," *FRP Compos. Civ. Eng.*, p. 267, 2008.
390. A. M. Mahmoud *et al.*, "Non-destructive ultrasonic evaluation of CFRPconcrete specimens subjected to accelerated aging conditions," *NDT E Int.*, vol. 43, no. 7, pp. 635–641, 2010, doi: 10.1016/j.ndteint.2010.06.008.
391. J. C. Trovillion, V. F. Godínez-Azcuaga, and R. D. Finlayson, "A nondestructive evaluation technique for Fiber Reinforced Polymer (Frp) composites using Acoustic Guided Waves (Agw)."
392. E. L. M. Ribolla, M. R. Hajidehi, G. F. Scimemi, A. Spada, and G. Giambanco, "Assessment of bonding defects in FRP reinforced structures via ultrasonic technique," *Chall. J. Struct. Mech.*, 2016, doi: 10.20528/cjsmec.2016.08.018.
393. P. Corvaglia and A. Largo, "IRT survey for the quality control of FRP reinforced r.c. structures," 2008.
394. N. Banthia, A. Aidin, M. Demers, A. Mufti, and S. Sheikh, "Durability of FRP-concrete bond in FRP-strengthened bridges," *Concr. Int.*, vol. 32, no. 8, pp. 45–51, 2010.
395. A. Kylili, P. A. Fokaides, P. Christou, and S. A. Kalogirou, "Infrared thermography (IRT) applications for building diagnostics: A review," *Appl. Energy*, vol. 134, pp. 531–549, 2014.
396. B. Wang, S. Zhong, T.-L. Lee, K. S. Fancey, and J. Mi, "Non-destructive testing and evaluation of composite materials/structures: A state-of-the-art review," *Adv. Mech. Eng.*, vol. 12, no. 4, p. 1687814020913761, 2020.
397. J. R. Brown and S. H. Chittineni, "Comparison of lock-in and pulse-phase thermography for defect characterization in FRP composites applied to concrete," *Thermosense Therm. Infrared Appl. XXXVII*, vol. 9485, p. 94850B, 2015, doi: 10.1117/12.2177260.
398. G. Zi, J. G. Sim, H. Oh, and J. Lee, "Optimum NDT using infrared thermography for defected concrete," in *Bridge Maintenance, Safety, Management, Health Monitoring and Informatics—IABMAS '08: Fourth International IABMAS Conference*, 2008, pp. 1223–1230.
399. J. Dumoulin, C. Ibarra-Castanedo, M. Quiertant, F. Taillade, A. Bendada, and X. Maldague, "Evaluation of FRP gluing on concrete structures by active infrared thermography by," 2010.
400. B. Milovanović and I. Banjad Pečur, "Review of active IR thermography for detection and characterization of defects in reinforced concrete," *J. Imaging*, vol. 2, no. 2, p. 11, 2016.

401. A. Foudazi, K. M. Donnell, and M. T. Ghasr, "Application of active microwave thermography to delamination detection," in *2014 IEEE International Instrumentation and Measurement Technology Conference (I2MTC) Proceedings*, 2014, pp. 1567–1571.
402. P. V. Mtenga, J. G. Parzych, and R. Limerick, "Quality assurance of FRP retrofit using infrared thermography," *Struct. - A Struct. Eng. Odyssey, Struct. 2001 - Proc. 2001 Struct. Congr. Expo.*, vol. 109, 2004, doi: 10.1061/40558(2001)4.
403. J. P. Nokes and G. F. Hawkins, "Infrared inspection of composite-reinforced concrete structures," 2001.
404. J. Tashan and R. Al-Mahaidi, "Bond defect detection using PTT IRT in concrete structures strengthened with different CFRP systems," *Compos. Struct.*, vol. 111, no. 1, pp. 13–19, 2014, doi: 10.1016/j.compstruct.2013.12.016.
405. B. Wen *et al.*, "Pulse-heating infrared thermography inspection of bonding defects on carbon fiber reinforced polymer composites," *Sci. Prog.*, vol. 103, no. 3, Jul. 2020, doi: 10.1177/0036850420950131.
406. L. Cantini, M. Cucchi, G. Fava, and C. Poggi, "Damage and defect detection through infrared thermography of fiber composites applications for strengthening of structural elements," *RILEM Bookseries*, vol. 6, pp. 779–784, 2012, doi: 10.1007/978-94-007-0723-8_111.
407. M. M. Caldeira and I. J. Padaratz, "Potentialities of infrared thermography to assess damage in bonding between concrete and GFRP," *Rev. IBRACON Estruturas e Mater.*, vol. 8, no. 3, pp. 296–322, Jun. 2015, doi: 10.1590/s1983-41952015000300004.
408. F. Taillade, M. Quiertant, K. Benzarti, and C. Aubagnac, "Shearography and pulsed stimulated infrared thermography applied to a nondestructive evaluation of FRP strengthening systems bonded on concrete structures," *Constr. Build. Mater.*, vol. 25, no. 2, pp. 568–574, 2011, doi: 10.1016/j.conbuildmat.2010.02.019.
409. W. L. Lai, S. C. Kou, C. S. Poon, W. F. Tsang, S. P. Ng, and Y. Y. Hung, "Characterization of flaws embedded in externally bonded CFRP on concrete beams by infrared thermography and shearography," *J. Nondestruct. Eval.*, vol. 28, no. 1, pp. 27–35, 2009, doi: 10.1007/s10921-009-0044-x.
410. U. B. Halabe, S. S. Dutta, and H. V. S. GangaRao, *NDE of FRP Wrapped Columns Using Infrared Thermography*. American Institute of Physics, 2008.
411. M. R. Valluzzi, E. Grinzato, C. Pellegrino, and C. Modena, "IR thermography for interface analysis of FRP laminates externally bonded to RC beams," *Mater. Struct. Constr.*, vol. 42, no. 1, pp. 25–34, Jan. 2009, doi: 10.1617/s11527-008-9364-z.
412. K. K. Ghosh and V. M. Karbhari, "Use of infrared thermography for quantitative nondestructive evaluation in FRP strengthened bridge systems," *Mater. Struct. Constr.*, vol. 44, no. 1, pp. 169–185, Jan. 2011, doi: 10.1617/s11527-010-9617-5.
413. J. Tashan, R. Al-Mahaidi, and A. Mamkak, "Defect size measurement and far distance infrared detection in CFRP-concrete and CFRP-steel systems," *Aust. J. Struct. Eng.*, vol. 17, no. 1, pp. 2–13, Jul. 2016, doi: 10.1080/13287982.2015.1116177.
414. N. J. Mabry, K. J. Peters, and R. Seracino, "Depth detection of bond defects in multilayered externally bonded CFRP-to-Concrete using pulse phase thermography," *J. Compos. Constr.*, vol. 19, no. 6, p. 04015002, 2015, doi: 10.1061/(asce)cc.1943-5614.0000551.
415. J. C. Gu, S. Unjoh, and H. Naito, "Detectability of delamination regions using infrared thermography in concrete members strengthened by CFRP jacketing," *Compos. Struct.*, vol. 245, Aug. 2020, doi: 10.1016/j.compstruct.2020.112328.
416. C. Maierhofer, A. Brink, M. R€ O Ollig, and H. Wiggenhauser, "Transient thermography for structural investigation of concrete and composites in the near surface region," *Infrared Phys. Technol.*, vol. 43, no. 3–5, pp. 271–278, 2002, [Online]. Available: www .elsevier.com/locate/infrared.

417. L.-D. Théroux, J. Dumoulin, and X. Maldague, "Square pulse heating infrared thermography and shearography applied simultaneously on CFRP tissue bonded to reinforced concrete," no. November 2018, 2014, doi: 10.21611/qirt.2014.209.
418. T. Yu, T. K. Cheng, A. Zhou, and D. Lau, "Remote defect detection of FRP-bonded concrete system using acoustic-laser and imaging radar techniques," *Constr. Build. Mater.*, vol. 109, pp. 146–155, 2016, doi: 10.1016/j.conbuildmat.2015.12.113.
419. Q. Qiu and D. Lau, "Defect detection of FRP-bonded civil structures under vehicle-induced airborne noise," *Mech. Syst. Signal Process.*, vol. 146, Jan. 2021, doi: 10.1016/j.ymssp.2020.106992.
420. A. Haque and P. K. Raju, "Sensitivity of the acoustic impact technique in characterizing defects/damage in laminated composites," *J. Reinf. Plast. Compos.*, vol. 14, no. 3, pp. 280–296, 1995.
421. M. Gower *et al.*, "Design and manufacture of reference and natural defect artefacts for the evaluation of NDE techniques for Fibre Reinforced Plastic (FRP) composites in energy applications," 2016. [Online]. Available: http://creativecommons.org/licenses/by/3.0/.
422. M. Aboukhousa and N. Qaddoumi, "Near-field microwave imaging of subsurface inclusions in laminated composite structures," 2004.
423. B. C. Ray, S. T. Hasan, and D. W. Clegg, "Evaluation of defects in FRP composites by NDT techniques," *J. Reinf. Plast. Compos.*, vol. 26, no. 12, pp. 1187–1192, 2007, doi: 10.1177/0731684407079348.
424. N. Kim and J. D. Achenbach, "Quantitative characterization of multiple delaminations in laminated composites using the Compton backscatter technique," 1998.
425. C. Scarponi and G. Briotti, "Ultrasonic technique for the evaluation of delaminations on CFRP, GFRP, KFRP composite materials," *Compos. Part B Eng.*, vol. 31, pp. 237–243, 2000.
426. K. S. Tan, N. Guo, B. S. Wong, and C. G. Tui, "Comparison of Lamb waves and pulse echo in detection of near-surface defects in laminate plates," *NDT E Int.*, vol. 28, no. 4, pp. 215–223, 1995, doi: 10.1016/0963-8695(95)00023-Q.
427. W. Hillger, M. Ahrholdt, H. Rohling, and R. Henrich, "Non-contact ultrasonic imaging techniques for composite components," 2004.
428. M. V. Hosur, U. K. Vaidya, C. Ulven, and S. Jeelani, "Performance of stitched/unstitched woven carbon/epoxy composites under high velocity impact loading," *Compos. Struct.*, vol. 64, no. 3–4, pp. 455–466, 2004, doi: 10.1016/j.compstruct.2003.09.046.
429. K. Imielińska, M. Castaings, R. Wojtyra, J. Haras, E. Le Clezio, and B. Hosten, "Air-coupled ultrasonic C-scan technique in impact response testing of carbon fibre and hybrid: Glass, carbon and Kevlar/epoxy composites," *J. Mater. Process. Technol.*, vol. 157–158, no. SPEC. ISS., pp. 513–522, 2004, doi: 10.1016/j.jmatprotec.2004.07.143.
430. W. Bai and B. S. Wong, "Evaluation of defects in composite plates under convective environments using lock-in thermography," *Meas. Sci. Technol.*, vol. 12, no. 2, pp. 142–150, 2001, doi: 10.1088/0957-0233/12/2/303.
431. M. Saeedifar, M. Fotouhi, M. Ahmadi Najafabadi, H. Hosseini Toudeshky, and G. Minak, "Prediction of quasi-static delamination onset and growth in laminated composites by acoustic emission," *Compos. Part B Eng.*, vol. 85, no. October 2017, pp. 113–122, 2016, doi: 10.1016/j.compositesb.2015.09.037.
432. S. W. Choi and J. H. Lee, "Nondestructive evaluation of internal defects for composite materials by using shearography," *Key Eng. Mater.*, vol. 270–273, no. I, pp. 781–786, 2004, doi: 10.4028/www.scientific.net/kem.270-273.781.
433. J. Weissenböck, A. Reh, D. Salaberger, C. Heinzl, and J. Kastner, "Advanced visualization and exploration techniques for fiber reinforced polymers," *11th Eur. Conf. Non-Destructive Test.*, no. Ecndt, pp. 1–10, 2014.

434. M. Rodríguez-Hortalá, J. Hatzmann, and H. P. Degischer, "Structural characterisation of defects in CFRP laminates produced with different epoxy resin systems and their influence on the mechanical properties," *ECCM 2012 – Compos. Venice, Proc. 15th Eur. Conf. Compos. Mater.*, no. June, pp. 24–28, 2012.
435. C. Ékes, "GPR: A new tool for structural health monitoring of infrastructure," 2007.
436. M. Ghaib, "Detection and localization of damage in fiber reinforced polymer bars using acoustic emission, micro computed tomography, and scanning electron microscopy techniques," no. August, p. 197, 2018.
437. A. Mehrabi, S. S. Khedmatgozar Dolati, P. Malla, A. Nanni, and J. D. Ortiz, "A framework for field inspection of in-service FRP reinforced / strengthened concrete bridge elements," 2024. [Online]. Available: https://rosap.ntl.bts.gov/view/dot/73333.
438. P. Böer, L. Holliday, and T. H. K. Kang, "Interaction of environmental factors on fiber-reinforced polymer composites and their inspection and maintenance: A review," *Constr. Build. Mater.*, vol. 50, pp. 209–218, 2014, doi: 10.1016/j.conbuildmat.2013.09.049.
439. S. S. Khedmatgozar Dolati, P. Malla, J. D. Ortiz, A. Mehrabi, and A. Nanni, "Non-destructive testing applications for in-service FRP reinforced/strengthened concrete bridge elements," in *Nondestructive Characterization and Monitoring of Advanced Materials, Aerospace, Civil Infrastructure, and Transportation XVI*, 2022, vol. 12047, pp. 59–74.
440. M. M. Gauthier, *Engineered Materials Handbook*, vol. 1. ASM International, 1995.
441. E. K. Gamstedt and B. A. Sjögren, "Micromechanisms in tension-compression fatigue of composite laminates containing transverse plies," *Compos. Sci. Technol.*, vol. 59, no. 2, pp. 167–178, 1999, doi: 10.1016/S0266-3538(98)00061-X.
442. D. T. G. Katerelos, J. Varna, and C. Galiotis, "Energy criterion for modelling damage evolution in cross-ply composite laminates," *Compos. Sci. Technol.*, vol. 68, no. 12, pp. 2318–2324, 2008, doi: 10.1016/j.compscitech.2007.09.014.
443. R. Talreja and C. V. Singh, *Damage and Failure of Composite Materials*. Cambridge, UK: Cambridge University Press, 2012.
444. ACI 440.2R-02, *Guide for the Design and Construction of Externally Bonded FRP Systems for Strengthening Concrete Structures (ACI 440.2R-02)*. 2005.
445. A. Nanni and A. Lopez, "Validation of FRP composite technology through field testing," 2004.
446. M. M. Fayyadh and H. Abdul Razak, "Assessment of effectiveness of CFRP repaired RC beams under different damage levels based on flexural stiffness," *Constr. Build. Mater.*, vol. 37, pp. 125–134, 2012, doi: 10.1016/j.conbuildmat.2012.07.021.
447. ACI 224.1R-07, *Causes, Evaluation, and Repair of Cracks in Concrete Structures*. 2007.
448. M. Yumnam, H. Gupta, D. Ghosh, and J. Jaganathan, "Inspection of concrete structures externally reinforced with FRP composites using active infrared thermography: A review," *Constr. Build. Mater.*, vol. 310, no. May, p. 125265, 2021, doi: 10.1016/j.conbuildmat.2021.125265.
449. H. N. Walker, "Petrographic methods of examining hardened concrete: A petrographic manual, Publication No. FHWA-RD-97-146," 1997. [Online]. Available: https://www.fhwa.dot.gov/publications/research/infrastructure/pavements/pccp/97146/chaptr06.cfm.
450. T. M. McCabe, "Quality control in concrete pavement: Early detection of near-surface honeycombing with GPR," University of Nebraska, 2020.
451. K. F. Barry *et al.*, "Void detection system," 2013.
452. M. L. Gambhir, *Concrete Technology*, 5th ed. New Delhi: McGraw-Hill Education (India), 2017.

453. S. Siddique, S. Chaudhary, S. Shrivastava, and T. Gupta, "Sustainable utilisation of ceramic waste in concrete: Exposure to adverse conditions," *J. Clean. Prod.*, vol. 210, pp. 246–255, 2019.
454. G. Tumialan, P. Serra, A. Nanni, and A. Belarbi, "Concrete cover delamination in reinforced concrete beams strengthened with carbon fiber reinforced polymer sheets," *Am. Concr. Institute, ACI Spec. Publ.*, vol. SP-188, pp. 725–735, 1999, doi: 10.14359/5667.
455. B. S. da Fonseca, A. S. Castela, M. A. G. Silva, R. G. Duarte, M. G. S. Ferreira, and M. F. Montemor, "Influence of GFRP confinement of reinforced concrete columns on the corrosion of reinforcing steel in a salt water environment," *J. Mater. Civ. Eng.*, vol. 27, no. 1, p. 04014107, 2015, doi: 10.1061/(asce)mt.1943-5533.0001039.
456. H. Liang, S. Li, Y. Lu, J. Hu, and Z. Liu, "Electrochemical performance of corroded reinforced concrete columns strengthened with fiber reinforced polymer," *Compos. Struct.*, vol. 207, no. August 2018, pp. 576–588, 2019, doi: 10.1016/j.compstruct.2018.09.028.
457. S. Masoud and K. Soudki, "Evaluation of corrosion activity in FRP repaired RC beams," *Cem. Concr. Compos.*, vol. 28, no. 10, pp. 969–977, 2006, doi: 10.1016/j.cemconcomp.2006.07.013.
458. R. Sen, "Durability of advanced composites in a marine environment," *Int. J. Mater. Prod. Technol.*, vol. 19, no. 1–2, pp. 118–129, 2003, doi: 10.1504/IJMPT.2003.003548.
459. W. Li, "Innovative structural health monitoring using smart sensors," University of Houston, 2017.
460. W. Li, S. Fan, S. C. M. Ho, J. Wu, and G. Song, "Interfacial debonding detection in fiber-reinforced polymer rebar–reinforced concrete using electro-mechanical impedance technique," *Struct. Heal. Monit.*, vol. 17, no. 3, pp. 461–471, 2018, doi: 10.1177/1475921717703053.
461. T. Kanakubo, K. Yonemaru, H. Fukuyama, M. Fujisawa, and Y. Sonobe, "Bond performance of concrete members reinforced with FRP bars," *ACI Spec. Publ.*, vol. 138, pp. 767–788, 1993.
462. S. C. M. Ho *et al.*, "Inference of bond slip in prestressed tendons in concrete bridge girders," *Struct. Control Heal. Monit.*, 2015, doi: 10.1002/stc.
463. B. Di, J. Wang, H. Li, J. Zheng, Y. Zheng, and G. Song, "Investigation of bonding behavior of FRP and steel bars in self-compacting concrete structures using acoustic emission method," *Sensors (Switzerland)*, vol. 19, no. 1, pp. 1–14, 2019, doi: 10.3390/s19010159.
464. L. Huo, H. Cheng, Q. Kong, and X. Chen, "Bond-slip monitoring of concrete structures using smart sensors—A review," *Sensors (Switzerland)*, vol. 19, no. 5, 2019, doi: 10.3390/s19051231.
465. S. Valentine, "Expected service life of michigan department of transportation reinforced concrete bridge decks," 2015.
466. A. Nanni, A. De Luca, and H. Zadeh, *Reinforced Concrete with FRP Bars: Mechanics and Design*. Boca Raton, FL: CRC Press, 2014.
467. W. K. Feeser and V. L. Brown, "Guide examples for design of concrete reinforced with FRP bars," *Am. Concr. Institute, ACI Spec. Publ.*, vol. SP-230, pp. 935–954, 2005.
468. S. S. K. Dolati, P. Malla, J. D. Ortiz, A. Mehrabi, and A. Nanni, "Identifying NDT methods for damage detection in concrete elements reinforced or strengthened with FRP," *Eng. Struct.*, vol. 287, p. 116155, 2023.
469. A. Baniya, S. Milev, R. Henry, and J. Tatar, "Durability assessment of externally bonded Fiber Reinforced Polymer (FRP) composite repairs in bridge [Project]," 2022.
470. J. M. Pevey, W. B. Rich, C. S. Williams, and R. J. Frosch, "Repair and strengthening of bridges in Indiana using Fiber Reinforced Polymer systems: Volume 1—Review of current FRP repair systems and application methodologies," 2021. doi: 10.5703/1288284317309.

471. J. Wang and Z. Fan, "Detecting of the crack and leakage in the joint of precast concrete segmental bridge using piezoceramic based smart aggregate," *Sensors (Switzerland)*, vol. 20, no. 18, pp. 1–17, 2020, doi: 10.3390/s20185398.
472. B. A. Graybeal, "Flexural capacity of fire-damaged prestressed concrete box beams," no. February, 2007, [Online]. Available: http://trid.trb.org/view.aspx?id=806369.
473. E. El-Salakawy, C. Kassem, and B. Benmokrane, "Field application of FRP composite bars as reinforcement for bridge decks," in *4th Structural Specialty Conference of the Canadian Society for Civil Engineering*, 2002, pp. 1–10.
474. M. Abedin, F. J. De Caso y Basalo, N. Kiani, A. B. Mehrabi, and A. Nanni, "Bridge load testing and damage evaluation using model updating method," *Eng. Struct.*, vol. 252, no. September 2021, p. 113648, 2022, doi: 10.1016/j.engstruct.2021.113648.
475. AASHTO, "Manual for bridge element inspection." American Association of State Highway and Transportation Officials, Washington, DC., 2019.
476. FHWA, "Specifications for the National Bridge Inventory," 2022.
477. A. Baniya, S. Milev, R. Henry, and J. Tatar, "Durability assessment of externally bonded Fiber Reinforced Polymer (FRP) composite Repairs in bridges," 2022, [Online]. Available: https://trid.trb.org/view/1687435.
478. Emergen Research, "Fiber reinforced polymer rebar market," 2022. [Online]. Available: https://www.emergenresearch.com/press-release/global-fiber-reinforced-polymer-rebar-market.
479. B. Phares, Z. Liu, and K. Freeseman, "Advancing bridge load rating: State of practice and frameworks," 2022.
480. AASHTO, *The Manual for Bridge Evaluation*, 3rd ed. Washington, DC: American Association of State Highway and Transportation Officials, 2018.
481. NBIS, *National Bridge Inspection Standards [NBIS]*. 2022, pp. 1–20.
482. US Department of Defense, *Polymer Matrix Composites: Materials Usage, Design, and Analysis: CMH-17 Volume 3 - Revision G*. SAE International, 2012.
483. M. S. Lenett, A. J. Helmicki, and V. J. Hunt, "Multi-reference impact testing of FRP bridge deck material," 2000.
484. C. Dumlao *et al.*, "Demonstration low-cost modular composite highway bridge," 1996.
485. V. M. Karbhari, J. W. Chin, and D. Reynaud, "Critical gaps in durability data for FRP composites in civil infrastructure," *International SAMPE Symposium and Exhibition (Proceedings)*, vol. 45. Long Beach, CA, 2000.
486. P. K. Dutta, S.-C. Kwon, and R. Lopez-Anido, "Fatigue performance evaluation of FRP composite bridge deck protypes under high and low temperatures," 2003.
487. J. L. Senne, S. W. Case, and T. E. Cousins, "Fatigue life of FRP composite beams," Virginia Polytechnic Institute and State University, 2000.
488. H. C. Tang, T. Nguyen, T. Chuang, J. Chin, J. Lesko, and H. F. Wu, "Fatigue model for fiber-reinforced polymeric composites," *J. Mater. Civ. Eng.*, vol. 12, no. 2, pp. 97–104, 2000.
489. T. Eljufout, "A comparative study on hybrid fatigue stress-life model of RC beams strengthened with NSM CFRP," *Mech. Based Des. Struct. Mach.*, vol. 51, no. 12, pp. 6814–6827, 2023, doi: 10.1080/15397734.2022.2068577.
490. J. Deng *et al.*, "Fatigue behaviour of CFRP bar-reinforced seawater sea sand concrete beams: Deformation analysis and prediction," *Buildings*, vol. 13, no. 9, 2023, doi: 10.3390/buildings13092273.
491. H. P. Kan, "Enhanced reliability prediction methodology for impact damaged composite structures," 1998.

492. D. McGovan and D. Ambur, "Damage characteristics and residual strength of composite sandwich panels impacted with and without a compression loading," in *39th AIAA/ASME/ASCE/AHS/ASC Structures, Structural Dynamics, and Materials Conference and Exhibit*, 1998, p. 1783.
493. T. Nyman, "Fatigue and residual strength of composite aircraft structures, report 99-26," 1999.
494. J. Tomblin, T. Lacy, B. Smith, and S. Hooper, "Review of damage tolerance for composite sandwich airframe structures, DOT/FAA/AR-99/49," Washington DC, 1999.
495. J. S. Tomblin, K. S. Raju, J. F. Acosta, B. L. Smith, and N. A. Romine, "Impact damage characterization and damage tolerance of composite sandwich airframe structures – Phase II, DOT/FAA/AR-02/80," 2002.
496. M. C. Leong, M. Fotouhi, M. Jalalvand, X. C. Sun, and M. R. Wisnom, "A comparison of damage mechanisms in scaled static indentation tests on glass and carbon fibre laminates," 2017.
497. S. Case, J. South, J. Duthoit, and K. Reifsnider, "Life prediction techniques for composite materials," 1999.
498. S. Case and K. Reifsnider, "Simulation of performance and life prediction for composite laminates: MRLife12," 1999.
499. C. Nelon, O. Myers, and A. Hall, "The intersection of damage evaluation of fiber-reinforced composite materials with machine learning: A review," *J. Compos. Mater.*, vol. 56, no. 9, pp. 1417–1452, 2022, doi: 10.1177/00219983211037048.
500. D. Swartz and L. Ilcewitcz, "Fatigue and damage tolerance perspective for composite aircraft structures," in *6th Joint FAA/DOD/NASA Aging Aircraft Conference San Francisco, CA, USA September*, 2002, pp. 16–19.
501. P. Jolma, "Residual strength of laterally loaded CFRP/PVC sandwich panels with impact damage," 2001.
502. E. J. Barbero and L. De Vivo, "A constitutive model for elastic damage in fiber-reinforced PMC laminae," *Int. J. Damage Mech.*, vol. 10, no. 1, pp. 73–93, 2001.
503. S. W. Park, D. R. Veazie, and M. Zhou, "Post-impact aging of FRP composite laminates in the marine environment," 2000.
504. ACI 440.5-08, *Specification for Construction with Fiber-Reinforced Polymer Reinforcing Bars*. 2008.
505. Canadian Standards Association, *Specification for Fibre-Reinforced Polymers (FRP) (CSA S807-19)*. Toronto, ON: Canadian Standards Association, 2019.

# Index

## D

## L



## M



## N



## O



## P

For Product Safety Concerns and Information please contact our EU representative GPSR@taylorandfrancis.com Taylor & Francis Verlag GmbH, Kaufingerstraße 24, 80331 München, Germany

**Batch number: 10748235**

Printed by Printforce, the Netherlands